Vermiresource Technology

VERMIRESOURCE TECHNOLOGY

Edited By
G. Tripathi
Department of Zoology,
J.N.V. University,
Jodhpur–342 001

2003

DISCOVERY PUBLISHING HOUSE
NEW DELHI

First Published–2003

Reprinted - 2014

ISBN 978-81-7141-717-9

Published by:

DISCOVERY PUBLISHING HOUSE

4831/24, Prahlad Street, Ansari Road, Darya Ganj
New Delhi–110 002 (India)
Phone: 23279245, • Fax: 91-11-23253475
e-mail: dphtemp@indiatimes.com

Printed at :

Dynamic Printers, Delhi

Preface

Biological resources are the mirror of the past, property of the present and treasurer of the future. Rapid biological and biotechnological advancements are expected to change the complexion of the living world in the present millennium. It is the reason why we have started seeing bioresources and economics in an integrative way. Bioresources are treasurer of many unexplored genes that could be of a great use to mankind. Still we are allowing them to be destroyed even without identifying the species. A single important and useful biological species may be worth large sums of money. Among faunal resources, no animal is as useful as earthworms are for mankind and his environment. The earthworms and other vermiresources have enormous and diversified potentials for waste decomposition, biofertilizer and biopesticide production, land reclaimation and ecorestoration, environmental detoxification and protein generation. They may be used as food, drug and vitamins source. Recently, developed countries are conducting hightech vermiresource studies to explore usefulness of living worms. In fact vermiresource technology is an ecologically sound, economically viable and socially acceptable technology of major importance to agriculture and environment. Therefore, the present small book has been compiled to popularize the vermiresource technology among agricultural, environmental and biological scientists, industrialists and technologists, and administrators and policymakers etc. Students pursuing advance courses in science and technology so also literate people may show their interests in the book.

I express my deep sense of gratitude to Professor Janardan Singh, Department of Entomology and Agricultural

Zoology, Banaras Hindu University, Varanasi and Dr. J.M. Julka, Emeritus Scientist, Zoological Survey of India (ZSI), Solan, who inspired and guided me to learn vermicology., It is their encouragements which brought this compendious book in the present form for the benefits of a wider audience. I am thankful to my all research students for their cooperation given to me as and when required.

I am grateful to the Department of Biotechnology (DBT) and Dr. Manju Sharma, Secretary, DBT, New Delhi, for encouraging work on vermiresources biodiversity and vermitechnology.

Jodhpur G. Tripathi

List of Contributors

G. Tripathi, Department of Zoology, J.N.V. University, Jodhpur-342 001.

P. Bhardwaj, Department of Zoology, J.N.V. University, Jodhpur-342 001.

V.B. Lal, Department of Zoology, Koshi College, Khagaria-851 205.

S.C. Talashilkar, Department of Agricultural Chemistry and Soil Science, Dr. B.S. Konkan Krishi Vidyapeeth, Dapoli–415 712.

R.G. Kadam, Department of Agricultural Chemistry and Soil Science, Dr. B.S. Konkan Krishi Vidyapeeth, Dapoli–415 712.

A.A. Todkari, Department of Agricultural Chemistry and Soil Science, Dr. B.S. Konkan Krishi Vidyapeeth, Dapoli–415 712.

P.V. Inamdar, Department of Agricultural Chemistry and Soil Science, Dr. B.S. Konkan Krishi Vidyapeeth, Dapoli–415 712.

R.V. Dhopavakar, Department of Agricultural Chemistry and Soil Science, Dr. B.S. Konkan Krishi Vidyapeeth, Dapoli–415 712.

Contents

Contents

1

Biodiversity of Vermiresources

G. Tripathi and Poonam Bhardwaj

Introduction

Vermiresources are mainly annelids useful for man and his environment. Earthworms are important vermiresources having simple, cylindrical, coelomate and segmented body characterized by presence of setae. They are hermaphrodite and deposit their eggs in cocoon and develop clitellum at maturity. There is no well-marked head but a preoral lobe called the prostomium is present. The earthworms vary in size, colour and behaviour. The soil, moisture content, salinity, temperature, the type of organic matter they like to feed and the depth to which they can go to the soil vary for different types of earthworms. They form a major component of the soil biota and together with a large number of other organisms constitute the soil community. They are known for inhabiting the earth's soil since the Precambrian dating about the 600 million years, bearing a silent witness to plant and animal evolution through the several landmarks and debacles of the evolutionary paradox. The report on the finding of fossilized earthworm cocoons lends creditability to their ancestry. The basic design of the earthworm has not changed much over these million of years and also does not vary much between the species.

Earthworms are omnivorous animals but often selective in their food habits. They derive their nutrition from organic material, living bacteria, fungi, diatoms, algae, protozoa, nematodes and decomposing animals. Though earthworms are generally called as saprophages, they can be classified based on their feeding habits into detritivores and geophages. Detritivores feed at or near the soil surface, mainly on plant litter or dead roots and other plant debris in the organic matter rich surface soil horizon or mammalian dung. These worms are classified as humus formers and comprise of epigeic and anecic forms. However, geophagus worms feed deeper beneath the surface, ingesting large quantities of organic rich soil. They are generally called as humus feeders and comprise of the endogeic worms. Therefore, on the basis of ecological strategies, earthworms are divided into epigeic, endogeic and anecic forms. The worms live in the soil surface of 3-10 cm and feed on organic matter like leaf litter or animal excrements are called epigeic. They are very active and have high regenerative capacity within a short period of time. Normally they are richly pigmented worms. The endogeic worms live deep in the soil from 10-30 cm. Most of the endogeic worms feed on the humic materials and mineral matter. They have very long life cycle with limited regeneration capacity and lightly pigmented. The anecic worms may go very deep into soil up to 60-90 cm and form complicated burrows for their movement. They plaster the burrows with their own excrement and the mucus secretions.

Earthworms have dynamic potentials and can do wonderful jobs for man and biosphere. The potentials of earthworms have been proved in decomposition of waste materials. Breakdown of organic waste can turn such a negative asset into a profit, by producing useful materials, and at the same time minimizing environmental pollution. The vermicomposted materials may be used as biofertilizers. They have potentials to replace chemical fertilizer to some extent and create a better environment for development and growth of useful organisms to fight pests and pathogens of plants. Developed countries are earning millions of dollars from vermicomposting programme. Earthworms are also used as protein rich source for animal feed because they have 70-80% protein on a dry mass basis. This protein is of a high quality and has a good balance of essential amino acids and is especially rich in lysine. The amino acid composition of earthworm is for superior to snail and fish meat. Earthworms are used as a protein rich food source for fishery, pigery and poultry industries. They also used as a bait for fish throughout the world. Earthworms are not only a source of protein but they have excellent range of vitamins. Among these, niacin and B12 are of significant value. Niacin is a valuable component of animal feeds.

The earthworms are highly resistant to many pesticides and heavy metals. They may overcome the effects of such toxic chemicals by increasing mucus secretion, restricting the movements and increasing the reproductive potential upto certain concentration levels. They have capacity to accumulate pesticides and heavy metals in their tissues. The studies have shown that the earthworm concentrate cadmium, cobalt, mercury, zinc and lead. The accumulation of toxic chemicals in earthworm tissue is ecologically important because they are the important component in the food chain of several species of birds and mammals. Since the earthworms have the ability to take up in their tissues a number of unwanted chemicals, they may be used as bioindicators of pollution. The bioaccumulation of toxicants in the tissues of earthworms depends on the soil properties, pH, calcium concentration and so on. Earthworms also play a significant role in management and reclaimation of degraded pedoecosystem. The introduction of earthworms in degraded land enriches the exhausted soil with nutrients. The success of soil or land reclaimation depends upon identification of proper species suitable for different agroclimatic zones of a country and occurrence of threshold number of earthworms in a particular zone. Such practices are already on the way to improve the quality of exhausted or degraded soil in some countries.

In addition to earthworms being compost manufacturer, protein producer, land reclaimer and environmental cleaner, they are also known to be associated with medicine to cure various human diseases since ancient time. The pastes or extracts of earthworms have been used in treating wounds, chronic boils, piles, sore throat, hernia, chronic cough and bronchitis diptheria, jaundice, facial paralysis and impotency.

According to tradition Chinese medicine, earthworms possess antipyretic, antispasmodic, diuretic and detoxic effects etc. It has been found that earthworms have strong antiasthmatic, antihypertensive and desensitive effects. They are used in folk medicines to treat pyorrhoea, smallpox and to enable mothers to nurse their children's. It has been suggested that earthworms possess, antipyretic activity which is due to archiclonic acid. There is an agreeing historical evidence of the value of earthworms in curing rheumatism. Even in these days the Chinese, the Japanese and the Indians are said to use earthworms in various fancy medicines. The folk stories in China relate that earthworm exhibits contraceptive capacity. The spermatocidal effects of certain earthworm species have also been demonstrated. Earthworms are being utilized in medicinal research to study the regenerating mechanisms, cancer and birth control at some universities in USA. Undoubtedly, no animal is as useful for mankind as earthworm is.

History

The history of biology is sprinkled with research on the connections between present day biogeography and ecology on one hand and earth history on the other. The sparse fossil records provide little palaentological information regarding the history and development of Oligochaeta. The biogeography and evolution of earthworms are obscured due to paucity of fossil records. Nevertheless, their origin has been inferred from studies on the distribution, ecology and comparative anatomy of the extant species. On the basis of their food and feeding habits, Stephenson (1930) believed that they appeared in the Cretaceous when dicotyledonous plants came into existence. Michaelsen (1903) and Arldt (1908) estimated the origin of earthworms much earlier during the upper Jurassic and upper Triassic periods, respectively. Sims (1980) assumed that ancestors of present day earthworms were wide spread in the undivided palaeocontinent of Pangaea, which was formed at end of the Palaeozoic. The division of Pangaea in the Triassic resulted in the separation of northern continent of the Laurasia and a large southern mass of Gondwana. These further broke up into a smaller land masses during the subsequent periods and moved apart carrying with them the precursors of the present day earthworms (Julka, 1988).

Earthworms have drawn the attention of philosophers and naturalists since ancient times because of their great importance in soil improvement. As early as 384-322 B.C. the famous Greek Philosopher Aristotle described earthworms as 'the intestine of earth'. However, the zoological nomenclature of earthworms was adopted by Carolus Linnaeus (1758) who listed 2 annelid species in the 10th edition of his famous book entitled *"Systema Naturae":* an oligochaete *(Lumbricus terrestris)* and a polychaete *(Lumbricus marinus).* Existence of diversity in earthworms was brought to light by Savigny (1826) during his explorations of fauna of Paris region in France. He described 20 species of Lumbricidae from the area. More than a century ago, Charles Darwin was first to bring earthworms to the attention of scientists and the general public. He noted the importance of earthworms in breaking down dead plant materials, recycling the nutrients they contain, and turning over soil. After 40 years of observation he described earthworms in his book *"The Formation of Vegetable Mould through the Action of Worms"* (Darwin, 1881). In this book, he expressed the opinion that 'earthworms have played a most important part in the history of world'.

The oligochaete systematist Michaelsen named an earthworm genus after Wegener in honour of Wegener's work towards an understanding of the distributions of earthworms (Michaelsen, 1933). The importance of contributions of Darwin to our knowledge of earthworms cannot be stressed enough till mid 19th century. A great upsurge of work on the morphology, histology and taxonomy of earthworms occurred in the late 19th and early 20th centuries. However, it was only in the last 25 years that interest in and research into ecology and biology of earthworms has peaked. Much of this work has been summarized by Edwards and Lofty (1977) in their book on the *Biology and Ecology of earthworms*, by Satchell (1983) in *Earthworm Ecology: From Darwin to Vermiculture* and by Lee (1985) in his book on *Earthworms*: *Their Ecology and Relationships with Soil and Land Use.*

Species Diversity

The Megadrili Oligochaeta with 20 families and approximately 3,600 species of earthworms are distributed all over the world. The distribution of earthworms in soil is influenced by several factors. Among these factors, soil texture and aeration, temperature, moisture, pH, inorganic salts, organic matter, reproductive potentials and dispersive power of the species are important. The temperate soils have high percentage of earthworms, which feed directly on organic matter and low percentage of humus feeders. In tropical countries the humus feeders predominate over the organic matter feeding worms. Well-aerated loose soils rich with organic matter support worms largely than the hard, clay and poorly aerated soils. Earthworms are very sensitive to hydrogen ion concentration. Most of the species of earthworms prefer soil with about neutral pH. They avoid drought and dry soils either by migrating to lower layers or by entering a stage of diapause. Majority of worms occur in soils where the moisture content ranges from 12-45%. Generally earthworms are more active in moist soil than dry soil. The kind and amount of food materials available in soil influence the size of earthworm population, species diversity, growth rate and cocoon production.

Some reports have been published recently on earthworm diversity. Reynolds (1994) prepared an article on the *Earthworms of the World*. It deals with global distribution, barriers to migration, habitat requirements and functions of earthworms in the soil. The composition of different species of earthworms in different soils has been studied by a number of workers (Van-Rhee, 1967, Norstrom and Rundgren, 1973; Satchell, 1983; Marinissen and Bosch, 1992; Doube *et al.*, 1994a; Muys and Granval, 1997).In the northern regions of Europe, Asia and North America the Lumbricidae are the earthworms of forest, grassland, sown pasture, cropland and gardens (Lee,1985). A number of species have been reported from different parts of Australia (Baker and Barrett, 1994). The species *Eisenia fetida, Eisenia andrei, Gemascolex spp., Pheretimoid spp. and Spenceriella* spp. were recorded from all over Australia. Whereas the species *Aporrectodea longa* and *Lumbricus castaneus* have been recorded from Tasmania. While *Allolobophora chlorotica, Dendrobaena attemsi* and *Eiseniella tetraedra* were recorded from both Tasmania and Victoria. However, the species *Octolasion tyrtaeum* was recorded from Victoria only. The earthworm biodiversity and its potentials have been described by various workers (Dash and Patra, 1977; Bouche, 1981; Edwards, 1983; Ismail, 1986; Julka and Senapati, 1987; Tripathi *et al.*, 1995; Singh,1997). Julka and Paliwal (1986) have given a vivid description on the distribution of earthworms in northern mountains, Indo-Gangetic plain and Deccan Peninsula of India.

So far 509 species belonging to 67 genera and 10 families have been recorded from Indian subcontinent, indicating a high degree of diversity in this region as compared to other areas (Julka, 1993a). These families are Acanthodrilidae (3 genera and 34 species), Almidae (1 genus and 4 species), Criodrilidae (1 genus and 1 species), Eudrilidae (1 genus and 1 species), Glossoscolecidae (1 genus and 1 species), Lumbricidae (8 genera and 16 species), Megascolecidae (14 genera and 193 species), Moniligastridae (4 genera and 98 species), Ocnerodrilidae (8 genera and 16 species) and Octochaetidae (26 genera and 145 species). Among these, 385 species have been reported from India alone. *Amynthas* and *Metaphire* have endemicity in Myanmar and Andaman and Nicobar Islands. However, they along with other pheretimoids like *Pithermera* and *Polypheretima* are peregrine in the Indian and Srilankan regions. Fauna of the Andaman and Nicobar Islands is more closely related to that of Myanmar and Malaysia than to the Indian mainland.

About 68% of known species of earthworms in the subcontinent belong to 10 endemic genera with the break up as *Drawida* (79), *Perionyx* (53), *Eutyphoeus* (43), *Megascolex* (33), *Amynthas* (33), *Plutellus* (32), *Metaphire* (26), *Hoplochaetella* (18), *Tonoscolex* (16), and *Octochaetona* (15). The remaining 57 genera are either monospecific or represented by less than 10 species. Excluding a few widely distributed species, 38 of the endemic genera are only found in India. The rest of 7 genera are distributed in other parts of the world (species of *Plutellus* and *Megascolex* from Australian region possibly are not congeneric with Indian species). The peregrine genera are distributed among 8 families. These families are Lumbricidae, Ocnerodrilidae, Megascolecidae, Acanthodrilidae, Eudrilidae, Glossoscolecidae, Criodrilidae and Octochaetidae. Successful colonization of peregrine species is mainly due to their tolerance to a wide range of ecological conditions and to some extent parthenogenetic mode of reproduction in most of them. The extent of colonization of lumbricids has become so extensive in the western Himalayas that they now dominate over the endemic species at several places.

A total 42 species have been recorded from 8 western Himalayan districts viz., Dehra Dun, Tehri, Pauri, Uttarkashi, Chamoli, Almora, Nainital and Pithoragarh. *Eutyphoeus nainianus* Michaelsen and *Perionyx nainianus* Michaelsen are at present known from their type locality i.e., Nainital. More species are found in Dehra Dun district (28 spp.), followed by Nainital (23 spp.). On the other hand 6 species have been recorded from the district of Uttarkashi which is of different terrain and less explored (Julka, 1995). Uttarakhand in the western Himalayan forms part of the endemic area of the genera *Perionyx* and *Eutyphoeus*. The earthworm fauna of Rajasthan has been reported from 4 families. These are Lumbricidae, Ocnerodrilidae, Octochaetidea and Megascolecidae (Julka, 1996). Total 12 species have been recorded from different districts of Rajasthan. Some species of earthworms have been reported from Chennai (Ismail, 1986). A comprehensive systematic account and ecological and biological observation on the earthworms of Orissa have been given by Julka and Senapati (1987). They have reported 30 species from different parts of Orissa and these species belong to the families Moniligastridae, Almidae, Megascolecidae, Ocnerodrilidae, Octochaetidae and Acanthodrilidae. A total of 11 species of earthworms are recorded from Varanasi, Mirzapur and Allahabad district of Uttar Pradesh (Singh, 1997). Original photographs of some of the earthworm species recorded from Rajasthan (India) are shown in plate 1 and 2.

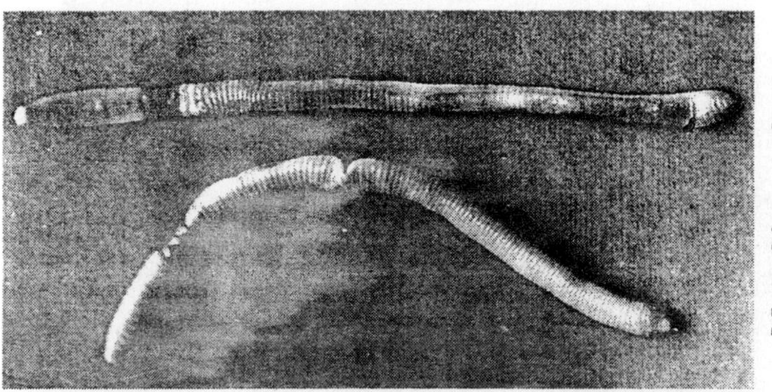

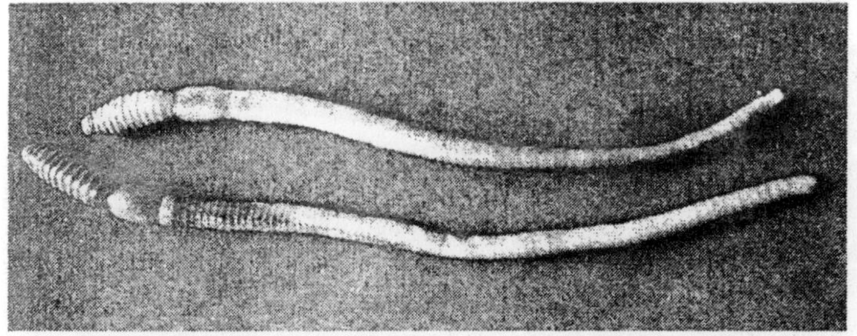

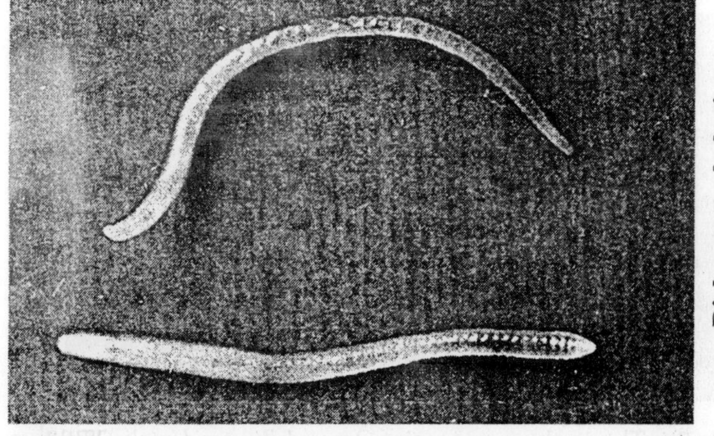

Plate 1: The earthworm species *Metaphire posthuma, Lampito mauritii* and *Dichogaster bolaui.*

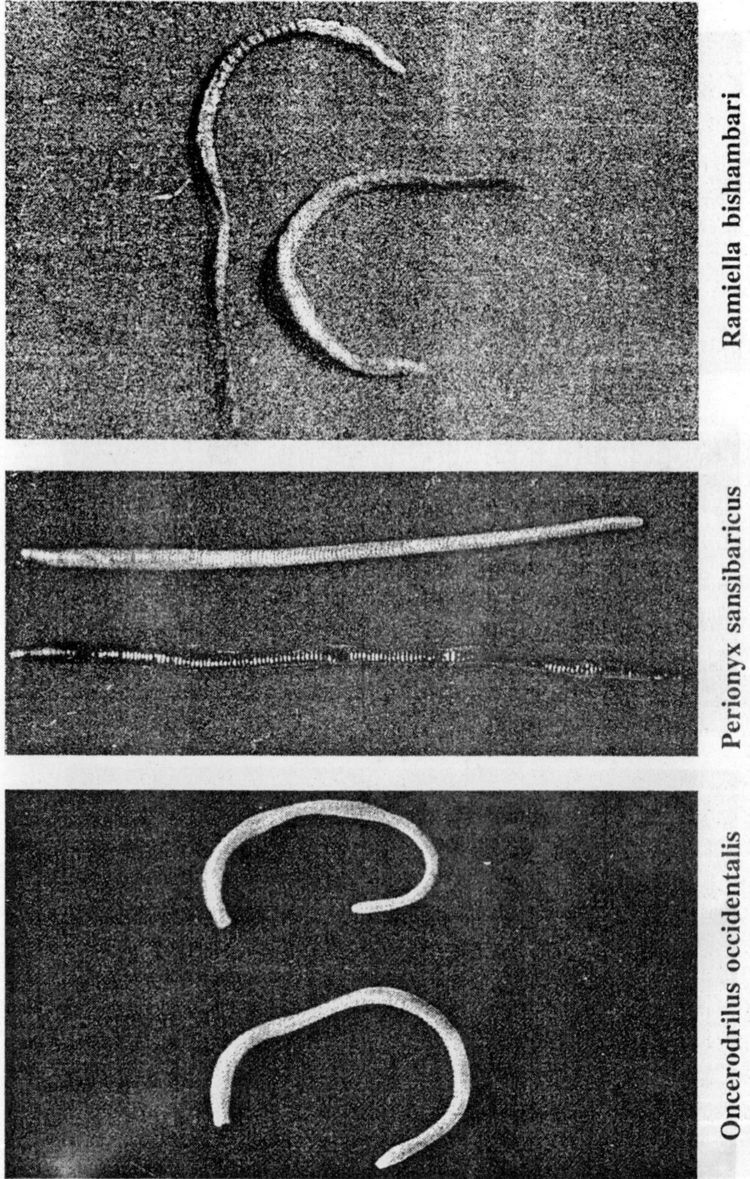

Plate 2: The earthworm species *Ocnerodrilus occidentalis, Perionyx sansibaricus* and *Ramiella bishambari*

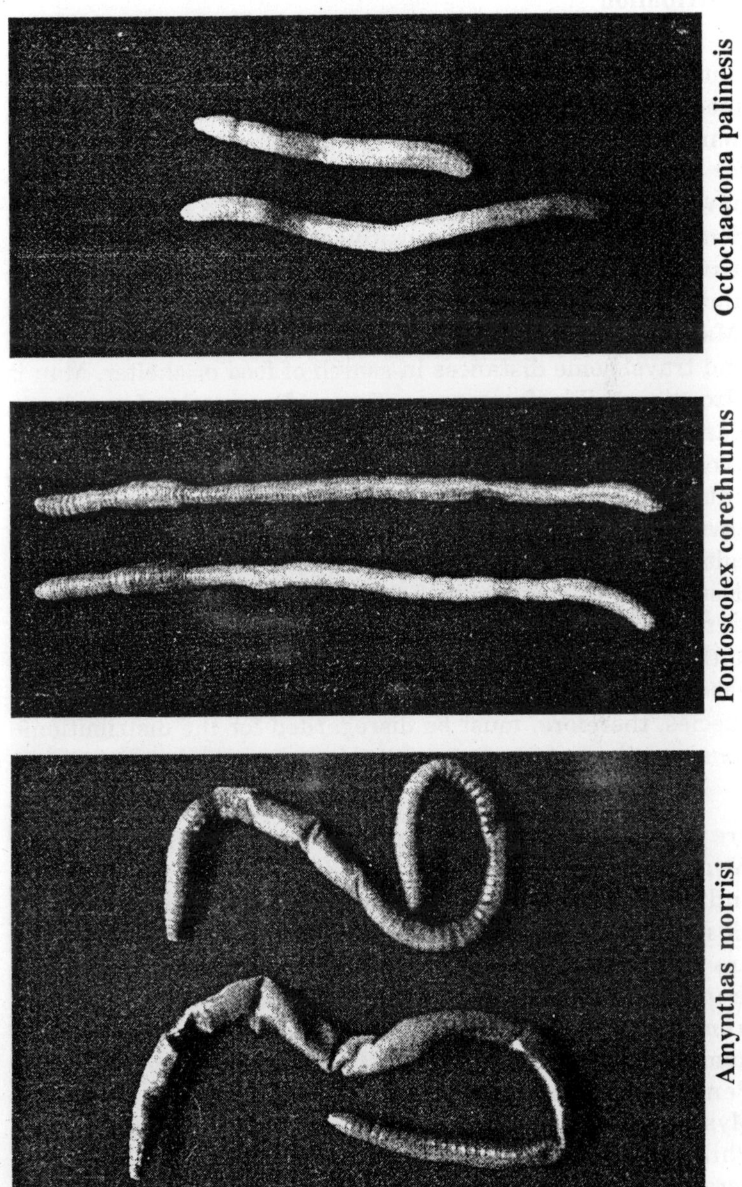

Plate 1: The earthworm species *Amynthas morrisis, Pontoscolex corethrurus* and *Octochaetona paliensis*

Distribution

Earthworms possess limited means of active migration. But the passive dispersal has distributed some earthworm species over wide areas. Most of them can not survive in marine environments, which are therefore, effective barriers for extension of their natural distribution. Endogeic species have, in general, a poor capacity to migrate (Bouche, 1983). Some of them are very sensitive to desiccation and do not leave the soil. Anecic worms are able to leave their burrows under certain conditions and colonize a few metres per annum (Mazaud and Bouche, 1980). Epigeic forms are most mobile and travel some distances in search of food or shelter. Man is also responsible for transporting a few species for culture. Earthworms are also transported unknowingly with the soil around roots of exotic plants. It is well known that anthropochorous species are able to colonize disturbed environments, which may be created by deforestation, intensive cultivation of new areas, pesticides, waste disposal, etc. Natural destruction of earthworm habitats may be due to glaciation, tectonic disturbance and changes in sea levels. Presence of anthropochorous and peregrine species in an area sometimes obscures the distributional pattern of earthworms. These species, therefore, must be disregarded for the distributional considerations of earthworms.

The earthworms belonging to the family Moniligastridae are primitive one with endemicity in eastern and southern Asia. Out of 5 genera known in the world, 4 are found in Indian subcontinent. These are *Desmogaster, Drawida, Hastirogaster* and *Moniligaster. Desmogaster* and *Hastirogaster* are restricted to Myanmar. A single record of *Desmogaster ferina* from Arunachal Pradesh may be due to its transportation from Myanmar. *Drawida* is the largest genus in terms of number of species. Its natural distribution extends to the Indian Peninsula, Eastern Himalayas and Northeast Ranges, Myanmar and as far as Korea, Japan, China, Manchuria, and Phillipine Islands, Sumatra and Borneo. Occurrence of peregrine species, *Drawida japonica* and *Drawida nepalensis*, in the Western Himalayas may be due to recent transportation. Moniligaster comprises a species, which are restricted in their distribution to the southern region of the Western Ghats (Julka, 1993b).The family Criodrilidae is mainly confined to the southwestern Palaearctic but some species have been carried to other regions, probably with water plants. Immature specimens of the widely distributed *Criodrilus lacuum* from our region belong to *Glyphidrilus* (Family Almidae).

The family Lumbricidae is endemic throughout the Palaearctic region and Eastern North America. Peregine lumbricids have acquired domicile in India at hill resorts with a temperate – like climate in the Himalayas, and Nilgiri and Palni Hills in the Peninsula. They are more widely distributed in Western Himalayas and form a dominant group in certain habitats at some places. In contrast, to Lumbricidae, the family Glossoscolocidae forms the dominant group in tropical South America. It is represented in the Indian subcontinent by a circummundane species, *Pontoscolex corethrurus* which is now widely distributed in the Indian Peninsula and northeast region. The family Almidae is represented by single genus, *Glyphidrilus* in Indian subcontinent. The *Glyphidrilus* spp. has endemicity in Myanmar, Northern Ranges, Ganga Plain, Western Ghats, Peninsular Plateaus and Sri Lanka. The genus is also indigenous to Africa, Malaysia and Indonesia. An African origin of the genus has been assumed. The family Ocnerodrilidae has two subfamilies viz., Ocnerodrilinae and Malabarinae. Members of Ocnerodrilinae in Indian subcontinent was possibly transported from Africa or South America where it is endemic. The members of Malabarinae occur mainly in Peninsular India, where as some species have their distribution into the Himalayan Hills and Myanmar.

The endemics of the family Acanthodrilidae are found in South America, some parts of North America, Africa, Southeast Asia and Australasia. Three genera occur in the Indian subcontinent. *Pontodrilus* and *Microscolex* are represented by the peregrines *P. bermudensis* and *M. phosphoreus*, respectively. *Plutellus* has endemicity mainly in the Western Ghats and Annamalai Hills in the Peninsula, Eastern Himalayas and Northeast Range and Myanmar. The family Octochaetidae occurs in New Zealand, Australia and spreads through the trophics of

America and Africa, including Medagascar, with the Asian representatives extending outward from Peninsular India. So far 26 Octochaetidae genera are known from the Indian subcontinent. Among these 25 are indigenous and form more than 50% of the known endemic genera of earthworms in the subcontinent (Julka, 1993b). However, the distributional range of the family Megascolecidae extends between the warm temperate Asia and Australasia. *Amynthas* and *Metaphire* are endemic in Myanmar and Andaman and Nicobar Islands, but are represented by peregrine species in other parts of the region. Both *Megascolex* and *Notoscolex* have also endemic species in Australasia. It is suspected that the Australasian species are not congeneric with the Indian forms.

Majority of earthworm genera in the Indian subcontinent have endemicity in the Deccan Peninsula (including Sri Lanka), Northeast India and Myanmar (Julka, 1993 b). The Peninsula was once a part of an ancient supercontinent Gondwana and has never been submerged under the sea. Excepting the Malayan Moniligastridae and Ethiopian Almidae, most of the indigenous earthworm genera presumably evolved and developed in the Deccan Peninsula. *Nelloscolex, Tonoscolex, Eutyphoeus* and *Scolioscolides* in the Northeast Region, and *Bahlia* in the Ganga Plains do not occur in the Peninsula. *Perionyx* and *Plutellus* are found in the Northeast Region and in the Peninsula and migrated to the Northeast across the Rajmahal gap. The species of *Amynthas* and *Metaphire* occurring in Myanmar have closer affinities with the corresponding Malayan and Chinese fauna. The species *Perionyx, Eutyphoeus* and *Glyphidrilus* have been recorded from isolated places in the Western Himalayas. It seems that the endemics in this region were exterminated during the last quaternary glaciation, which affected the area most than the other regions of the subcontinent (Wadia, 1973). Alternatively, the wave of westward migration of endemic earthworms from the east could not reach this region with full force (Julka, 1993b).

Biology

It is quite surprising that for the nearly 8,000 normal species of oligochaetes (Reynolds and Cook, 1993), modern, updated, life history studies have been made on only a few species, i.e., fewer than 20. Lee (1992) has suggested that only about six lumbricids and six tropical species have been studied in sufficient detail to provide adequate information. Mating is required in most species to exchange sperm from one individual to another. Fertilization is external in the cocoon, although internal fertilization has been reported in the Eudrilidae (Stephenson, 1930). Parthenogenesis is common in some species. In introduced species, parthenogenesis is considered to be of advantage, as a worm has not to find a mate to reproduce. Perhaps copulation in earthworms takes place due to pheromonal attraction between the two worms (Olive and Clark, 1978). It may occur above or below the surface of soil, and may last as long as an hour. Some species copulate periodically throughout the year except during unfavourable conditions, while in others mating is restricted to a particular period of the year (Julka, 1988).

After copulation cocoons are produced. These cocoons are deposited usually near the surface of the moist soil, but are led deeper if the soil is too dry. *Pheretima* sp. produces more cocoons from March to June and few during rainy months of July and August (Bahl, 1950). On the other hand, cocoons laid by *Octochaetona surensis, Lampito mauritii* and *Drawida willsi* increase in number from August and reach a peak density in October (Dash and Senapati, 1980). Each cocoon contains one to twenty fertilized ova but only a few, usually one or two, survive and hatch (Edwards and Lofty, 1977). However, there are reports of one to six young worms hatching from cocoons of *Eisenia fetida* and *Bimastos tumidus* (Vail, 1974). It is important to evaluate the number of live young earthworms that could be obtained from cocoons of each species. (Edwards, 1998) reported that *Eisenia fetida* produce 6 cocoons per worm per week (19 young worms), *Dendrobaena veneta* 5 cocoons (19 young worms), *Eudrilus eugeniae* 11 cocoons (20 young worms), *Perionyx excavatus* 24 cocoons (13 young worms), and *Perionyx hawayana* 10 cocoons (9 young worms) per parent worm. The time for cocoons to hatch varies from 13-126 days and time to reach the sexual maturity may be from 28 to 95 days. It suggests that the time required for an egg to reach its maturity may vary from 43 to 214 days. Senapati and Dash (1991) studied the life table of *Lampito mauritii* and *Octochaetona surensis*. They found that rate of cocoon production is higher in *L. mauritii* as compared to *O. surensis*.

The breeding season in earthworms occurred at different times within a year. Breeding season is known only for few Indian earthworm species. In the Gangetic Plain with a tropical climate, some peregrine species remain active throughout the year and become sexually mature as early as July and have a long breeding season. In contrast, some endemic species have restricted activity during the monsoons possess a short breeding season of 2-3 months (September-November) after the rains (Gates, 1945b). A peregrine lumbricid, *Octolasion tyrtaeum*, seems to breed in the Shimla Hills throughout the year, but the reproduction peak occurs in autumn and winter (Julka and Mukherjee, 1984). The activity of many species of earthworms is interrupted when the soil becomes too dry or soil temperature is either too low or too high. They become inactive and enter diapause or quiescent phase in order to tide over adverse environmental periods.

Categories and Richness

The observations on the life history and behaviour of common lumbricid species enabled Evans and Guild (1947,1948) and Guild (1951,1955) to distinguish deep dwelling forms from surface inhabiting ones. Graff (1953) distinguished lumbricids into two groups: (1) deep pigmented surface dwelling forms found primarily in habitats with sufficient organic matter and having high rate of reproduction, and (2) unpigmented forms inhabiting mineral horizon and with low reproduction rate. Svendsen (1957) documented the functional significance of pigmentation in earthworms. He suggested that red pigmentation in surface dwelling species provide protection from ultraviolet rays from sun. The New Zealand megascolecids have been divided by Lee into 'leaf mould' group, a 'top soil' group and a 'subsoil' group. In contrast, Wood (1974) divided Australian earthworms into 'topsoil' species, 'subsoil' species and other species (surface feeding/deep burrowing). A number of other workers have identified some important characteristic features to distinguish between surface dwelling and soil dwelling earthworms (Satchell, 1955; Stolte, 1962; Semenova,1968).

Perel (1977) classified lumbricids into two major groups on the basis of differences in their food habits: (i) 'humus formers', which feed upon slightly decomposed plant matter, and (ii) 'humus feeders', which feed upon advanced stage of decomposed plant material that has been incorporated into soil. Bouche (1977) proposed a generalized ecological classification for earthworms inhabiting different soil layers, usually into three groups: (i) surface or litter dwellers (epigeics), deeply pigmented and very active; (ii) top organs minerals soil dwellers (endogeics), which construct horizontal and branching burrows and are weakly pigmented and less active; and (iii). deep burrowers (anecics) which built vertical burrows but came to surface at night to feed or defecate, and are slightly pigmented at anterior and posterior ends. Lavelle (1983) has divided earthworms into five ecological categories viz., epigeic, anecic, oligo-, meso-, or polyhumic endogeic.

The occurrence of earthworm species at an individual site is generally low as compared to other biota in soil. Bornebusch (1930) observed 4-7 species in the forest of Denmark with the frequent occurrence of all the seven species of earthworms. Guild (1951) recorded 7-10 species in any one habitat in Scotland and showed the dominance of *Aporrectodea caliginosa* and *Aporrectodea longa*. Ljungstorm (1967) demonstrated the presence of 4-5 species in Swedish arable, coniferous and deciduous forest. Zajona (1968) found 2-9 species in arable location in Czechoslovakia where 4-5 species showed their common occurrence. El – Duweini and Ghabbour (1965) recorded 1-2 species from Egyptian location with common occurrence of single species. Three species have been reported from New Zealand Grasslands and Finlands coniferous forests (Lee 1959; Hahta *et al.*, 1967). Four species of earthworm have been reported from France and deciduous woodland in Canada with two showing normal occurrence (Bouche, 1969; Reynolds, 1972a). Singh (1997) reported the occurrence of 7-11 species from cultivated, non-cultivated grassland, gardenb and sewage soils from Uttar Pradesh in India and showed the common occurrence of 7 species in different pedoecosystems.

Few workers have sampled different habitats at same time of the year (Edwards and Lofty 1977). Therefore, it is not easy to assess the affect of varied habitats on the population densities of earthworms, especially when the sampling techniques employed are also different. Based on the studies of several European workers, low worm density was observed in mor soils, fallow soils and moorland than in mull soils, with variable numbers in regularly cultivated arable soils (Edwards and Lofty, 1977). Some of the density estimate for various Europeans habitats were: 18.5 – 33.5 m^{-2} in fallow soil (Drangaliev and Belousava 1969); 106 – 122m^{-2} in Oak woodland (Bornebusch 1930; Zazonc 1970); 73 – 177m^{-2} in beech woodland (Bornebusch 1930); and 68 – 157m^{-2} in mixed woodland (Reynoldson 1955; Edwards and Heath 1963); A low density of 14.2–142.4 m^{-2} was observed in a mixed woodland in the U.S.A. (Reynolds, 1972a). Various studies have been made on earthworm population densities in tropical soils. El – Duweini and Ghabbour (1965) estimated a density of 8 – 788m^{-2} in Egyptian soils as to compared to 7.4 – 101.8 m^{-2} in Ugandan soils (Block and Banage, 1968) and 33 m^{-2} in Nigerian grassland (Madge 1969).

Reynolds (1972a) explained that relative abundance and biomass of earthworms appeared to decrease with decline in soil moisture, soil acidity and palatability of food source. Reynolds (1972b) also showed that distribution of earthworms in Haliburton Highlights, Canada, was associated with some habitat factors. He reported that, about 80% worms occurs in surface soil between 14–18 C, 40% on each of the north and south facing slopes, 50% on terrace positions and 85% at places with surface soil acidity ranging between pH 5.0 – 6.9. According to Krishnamoorthy (1989), an increase in rainfall or soil moisture affect the worm populations. Chappell *et al.* (1971) and Aritajat *et al.* (1977) explained that density and biomass of earthworms decline as a result of trampling by human beings. Ismail *et al.*, (1990) also observed low values of earthworm densities (17.1m^{-2}) and biomass (1.47 gm m^{-2}) in locations affected by trampling.

Population densities of Indian earthworms have been reported for only a few ecosystems. The densities of worms are higher in pasture soils than woodlands, orchards, arable and barren soils. Dash and Patra (1979) recorded a density of 64 – 800 m^{-2} in a pasture in Orissa as compared to 25.1 - 936.5 m^{-2} in Karnataka (Kale and Krishnamoorthy, 1978). Reddy and Alfred (1978) observed a low density of 8.8 – 52 m^{-2} in a pine woodland in eastern India. *Drawida calebi*, *Lampito mauritii* and *Octochaetona surensis* are dominant in Orissan pastures (Senapati and Dash, 1981). Julka and Mukherjee (1984) studied the population dynamics of earthworms in western Himalaya. Ismail and Murthy (1985) recorded low population of earthworms from a site with a mat of undistributed organic matter on the soil surface. *Dichogaster bolaui* attains peak densities of 12617 m^{-2} in a compost pit and 8038 m^{-2} in a pasture at Sambalpur (Sahu *et al.*, 1988). In an orchard of Himachal Pradesh earthworm population attains a peak density of 362m^{-2} (Julka and Paliwal, 1990). In grassland of Andhra Pradesh *Drawida willsi* was the most dominant species (Reddy and Reddy, 1990).

Habitat Diversity

Earthworms are found in all types of soils provided there is sufficient moisture and food supply (Julka,1988a; Reynolds, 1973). They have been recorded from forests, grasslands, cultivated fields, orchards, gardens, plant nurseries and green houses (Gates, 1972; Julka, 1988a). A few occurrence of cavernicolous earthworms has been documented by Stephenson (1924a). A few species have been recorded to live under snow on high mountains (Julka, 1988a). Most earthworms inhabit neutral soils, but they are also found in acidic conditions (Olson, 1928; Bornebusch, 1930; Satchell, 1955). In Scotland, Guild (1948) observed preference of earthworms for light and medium loam soils than heavier clays, gravelly sands and alluvial soil. A few small sized species are able to survive in desert and semideserts (Kubiena, 1953; Kollmannsperger, 1956). Reynolds (1973) divided different habitats of earthworms into various categories such as (i) different types of soil including leaf mould, (ii) under logs and stones, under bark of dead and living trees,

(iii) in rotting logs and decaying organic matter accumulated in axils of tree branches, (iv) stream banks above water line (v) under stones embedded in streams, on banks of ponds and lakes, (vi) swamps, and (vii) intertidal zone on sea shores and brackish water adjacent to sea.

It has been reported that earthworms are not randomly distributed in soil (Guild, 1952). The pigmented lumbricids species *Dendrobaena octaedra*, *Lumbricus castaneus*, *Lumbricus rubellus* etc. aggregate under dung – pats (Boyd, 1958). The occurrence of earthworms is related with physicochemical conditions of the soil i.e., temperature, moisture, pH, C/N ratio etc. Most species of earthworms prefer soil with temperature of 10-35 C, moisture of 12-45%, pH about 7 and C/N ratio of 2-18 (Lavelle, 1974; Phillipson *et al.*, 1976; Edwards and Lofty, 1977; Kale and Krishnamoorthy, 1981; Lee, 1985; Govindan, 1998). Reinecke *et al.*, (1992) reported that the redworms (*Eisenia fetida*) can be cultivated in areas with higher temperature (as high as 43 C) as well as areas with lower soil temperatures (below 5 C). *Lumbricus terrestris* occurs in soil with a pH of 5.4 whereas *Megascolex* thrives well in soils with pH ranging from 4.5 to 4.7. *Eisenia fetida* dominates in soils of pH 7 to 8. *B. eiseni*, *D. exatedra* and *D. rutrida* are acid tolerant species and *A. caliginosa*, *A. nocturna*, *A. chlorotica*, *A. longa* and *A. rosea* are acid intolerant species. Soil pH may also influence the worms that go into diapause (Govindan, 1998).

Among these factors temperature seems to be important in determining individual metabolic rates. It can have major role in determining patterns of earthworm distribution and activity. Temperature may be a factor of primary importance in determining the composition and structure of earthworm communities (Lavelle, 1983; Lavelle *et al.*, 1989). High temperatures are often associated with moisture shortage. Seasonal earthworm mortality in temperate soils has been attributed to moisture stress rather than to temperature extremes (Gerad 1967; Phillipson, *et al.*, 1976). Earthworms are generally absent from very acidic soils (pH<3.5) and are scarce in soils with pH<4.5. Most of the temperate climate species are found in the range 5-7.4 (Satchell, 1967; Bouche, 1972). Other edaphic factors which have been linked with earthworm distribution include Ca, Mg and N content (Fragoso and Lavelle, 1992). While population can be adversely affected by high salt concentration which can occur for example in irrigated soils (El – Duweini and Ghabbour, 1965).

The species *A. caliginosa* survives well at moisture content varying from 15 – 34%. A moisture content of 23% favours *H. africans* to produce casts. Generally earthworms are more active in moist soil than dry soil. The abundance of *A. caliginosa* and *L. rubellus* has been correlated with soil temperature, moisture and supply of plant residues *Lampito mauritii* and *Pontoscolex corethrurus* are abundant in soils where C/N ratio is high but *Perionyx excavatus* is commonly found in soil with low C/N ratio (Govindan, 1998). *Lampito mauritii* has been reported to occur throughout the year at sites where annual temperature remains about 30 C, *Octochetona serrata* remains available at 27 C (Ismail and Murthy, 1985). The activity of *Lampito mauritii, Drawida calebi, Drawida willsi* and *Octochaetona surenesis* remains confined to 20 cm depth (Dash and Senapati, 1980).

Seasonal Dynamics

Several interdependent physicochemical factors affect the seasonal abundance and activities of earthworms (Edwards and Lofty, 1977). Among these soil temperature and soil moisture have been considered as most important (Reynolds and Jordon, 1975; Edwards and lofty 1977; Julka and Mukherjee, 1984). Evans and Guild (1947) found 5 lumbricids species as most active between August and September, and April to May in English pastures. However, the earthworms were most abundant in spring and autumn months in eastern United States with a humid continental climate (Hopp, 1947). Madge (1969) reported that the main activity of *Hyperiodrilus africanus* in Nigeria was in the beginning of wet season (May-June), followed by a gradual population decline until November. Surface dwelling species showed a single peak or prolonged emergence as compared to deep burrowing forms which have single or multiple peaks of emergence (Satchell, 1967;

Rundgren, 1977). Prashad (1916) firstly studied seasonal occurrence of earthworms in Indian subcontinent. He found that *Aporrectodea trapezoides* and *Allolobophora parva* were predominant in winter in Lahore. The population of *A. trapezoides* decline in spring. This species was not encountered in May, but the population of *Pheretima* spp. increased during this period. An introduced Lumbricid *Octolasion tyrtaeum* remain active throughout the year with peak density occurring during summer and winter rains in the mid hills of western Himalayas (Julka and Mukherjee, 1984). Kale and Krishnamoorthy (1982) observed seasonal changes and found a unimodal cycle in *Lampito maurttii* and bimodal cycles in *Pontoscolex corethrurus* and *Pheretima elongata* in Bangalore. Adult population of *Lampito mauritii* increased with total rainfall, but tropic factors were found to be more important.

Mishra and Ramakrishnan (1988) analysed population dynamics of earthworms in 3 successive fallows developed after slash and burn agriculture (Jhum cultivation) at an attitude of 960m in northeast India. They related monthly population fluctuations to soil temperature, moisture and litterfall pattern. The population biology and secondary production of *Dichogaster bolaui* have been noted by Sahu and Senapati (1986) in Orissa. *Dichogaster bolaui* showed a unimodal peak (8038 m^{-2}) at a pasture site during August and bimodal form of peak at a compost pit site with 12617 m^{-2} during August and 6538 m^{-2} during December (Sahu *et al.*, 1988). Its population was zero during summer due to adverse environmental conditions. Population dynamics of earthworms was investigated in northeast India by Bhadauria and Ramakrishnan (1989). The population densities of *Tonoscolex horai, Nelloscolex strigosus* and *Drawida assamensis* remain maximum in summer, whereas the density of *Amynthas corticis* reached its peak in winter. A relationship has been established between the population size and environmental parameters like soil moisture, temperature and organic matter.

The seasonal abundance of earthworms is affected by many ecological factors. In Indian subtropical climate, especially in the plains, earthworms are mainly active and abundant during the summer rains with maximum density occurring in September-October and minimum density in May-June (Gates, 1961b; Dash and Patra, 1977; Chauhan 1980). In a pine forests in Meghalaya, the maximum and minimum densities occurred in July and April, respectively (Reddy and Alfred, 1978). Ismail (1986) investigated that the highest and lowest number of earthworms in Chennai varied in different months with respect to different localities. The seasonal dynamics in grassland soils has been studied by various workers (Nakamura, 1968; Stein *et al.*, 1992). The annual fluctuations in the population of some tropical earthworms have been worked out in pasture soils (Senapati *et al.*, 1979). The seasonal changes in the population of earthworms have also been investigated in cultivated lands (Ghabbour and Shakir, 1982; Pani and Senapati, 1986). A statistically non-significant correlation between seasonal worm density and soil moisture (r=0.346), and soil temperature (r=0.529) was reported by Julka and Paliwal (1990).

Soil Enrichment

There has been an exponential increase in research addressing the impact of earthworms on terrestrial nutrient dynamics. Earthworms have been shown to affect key soil properties and processes i.e., microbial biomass and activities, organic matter dynamics, nutrient availability, plant uptake and production, soil structure and so on. There is a sufficient evidence to conclude that the earthworms should be considered key stone organisms in regulating nutrient cycling processes in many ecosystems. Earthworms may alter the balance between ecosystem conservation and loss of nutrients, particularly carbon (C) and nitrogen (N). Virtually, there is a need to concentrate on research that integrates soil biology, chemistry and physics. It may demonstrate how earthworms affect soil nutrient dynamics. Lavelle *et al.* (1998) highlighted the roles of earthworms in nutrient cycling. They stressed the need to determine the quality and quantity of soil organic matter ingested by earthworms because they may alter the ratio of label to recalcitrant organic matter and influence long term storage of soil C and N.

The changes in soil properties occur when earthworms colonize new soil. Invasion of earthworms into previously uncolonized soil alter microbial biomass, plant biomass and leaching of nutrients (Scheu and Parkinson, 1994 a,b) as well as soil C content and soil profile development (Alban and Berry, 1994). Gilot (1997) reported the effects of earthworms on farm production and soil characteristics in Africa. Earthworm changes soil structure by decreasing the percentage of small aggregates, and increasing the percentage of larger aggregates. These larger aggregates, primarily ageing costs, protect soil carbon as evidenced by decrease of 5% in carbon mineralization after 3 years. Devliegher and Verstraete (1997) believed that nutrient enrichment process (NEP) and gut associated process (GAP) of earthworms have contrasting effects on soil microbiology, chemistry and plant growth. For instance, *Lumbricus terrestris* incorporates and mixes surface organic matter with soil and increases biological activity and nutrient availability (NEP). However, they also assimilate nutrients from soil and organic matter as these materials pass through the gut (GAP). Bouche *et al.* (1997) developed models to quantify the direct role of earthworm in N cycle. Steinberg *et al.* (1997) examined the importance of earthworms in increasing the availability of N in forests.

At the ecosystem level, the fate of the N made available to the earthworm is of interest. Some of these N is available for plant uptake, because it leaches below the rooting zone. Blair *et al.* (1997) found an increase in concentration of soil nitrate deeper in the soil (30 – 45 cm) in earthworm addition treatments. Similarly, Subler *et al.* (1997) reported higher volume of Leachate at 45 cm in earthworm addition plots, and that the Leachate has much higher concentration of dissolved organic nitrogen (DON), compared to control plots. Norgrove and Hauser (2000) demonstrated that earthworm communities adjust the lower soil nutrient concentration by increasing their selectivity and thus produce relatively high casts on poorer soil. Increase in earthworm population by fallows lead to an increase in leaf litter decomposition, soil organic matter available phosphorous, extractable cations and pH (Tian *et al.*, 2000). Scullion and Malik (2000) illustrated the influence of earthworm activity on aggregate and organic matter stabilization. It has been observed that plant N capture increases from the sites of earthworm patches in the soil (Hodge *et al.*, 2000). In addition Edwards and Shipitalo (1998) have described the consequences of earthworms in agricultural soils emphasizing aggregation and porosity of soil. Kretzschmar (1998) reviewed the earthworm interaction with soil organization. A long term effects of earthworms on soil organic matter dynamics may vary depending the scale of time considered. When earthworms are introduced artificially into an ecosystem, they use part of carbon sources for their activity. They participate in accumulation of organic matter through an increase of organic matter produced in ecosystem and to the production of soil organic matter in structures of drilosphere (Martin, 1991). In general increase in nutrient turnover from earthworm activities usually results in increased plant growth (Lavelle *et al.*, 1998).

After 3 decades of unquestioned success, agriculture is known to face serious problems all over the world. In developed countries, huge increases in productivity have accompanied a severe depletion of soil quality in terms of resistance to erosion, organic contents and concentration of heavy metals and pesticide residues. In developing countries, modern agricultural practices are being intensified which are deteriorating the quality of soils, depleting the stocks of organic matter and eroding upper fertile layer of soil. The modern agricultural practices are on the way and the traditional practices are unable to preserve the soil fertility. According to Eswaran (1994) this degradation of soils causes many social and environmental problems. A common feature to all sorts of soil degradation is the significant decrease of organic reserves and a severe depletion of soil invertebrate communities, especially earthworms (Lavelle *et al.*, 1994). Saetre (1998) examined the role of *A.caliginosa* on carbon and nitrogen mineralization in soil. Earthworms may be considered as an important biological resource for the forming system, and the management of their communities is a promising field for progress in agricultural practices. Demand for techniques making optimal use of earthworm bioresources is likely to increase, but basic research is still needed to support their development.

Biodiversity Improvement

Interactions of the soil fauna, between both different faunal groups and soil microorganisms, are considered to be important in affecting soil processes and in influencing the community composition of various groups of the soil biota (Seastedt, 1984; Visser, 1985). Earthworms form an important group of the soil fauna, and show different ecological strategies (Doube and Brown, 1998). All facets of their life in organic and mineral soil layers can potentially have major impacts on other groups of soil organisms. Since earthworm species differ in size and behaviour, their activities have great consequences for the physical and chemical characteristics of soil (Lee and Foster, 1991). The activities include ingestion of soil and organic material and the intermixing of materials, ejection of gut contents as casts, and the formation of burrow systems (Guild 1955). Burrowing activities vary from species to species. These activities have a large impacts on microbial and faunal communities. There is a drilosphere soil zone around earthworm burrows (Bouche, 1975), which is generally rich in nitrogen, phosphorous and humified organic matter than in the surrounding soils (Beare *et al.,* 1994). Apart form burrowing, earthworms have significant effects on the fabric of mineral soil and surface organic layers.

Kubiena (1955) described that, in mineral soils possessing sufficient water-stable binding substances, worms can create a 'spongy fabric'. This fabric has good aeration and water status which enhances the development of aerobic bacteria, actinomycetes, and fungi. In compact soil fabrics, root development and earthworm activity may allow the development of 'channel fabrics'. This affects the soil structure, enhancing microbial and faunal development (Parkinson and Mclean, 1998). The impacts of earthworm on soil processes and on biomass dynamics of some major groups of soil biota have been described by Haimi and Huhta (1990); Haimi and Boucelhom (1991) and Scheu and Parkinson (1994). In classifying forest and healthland humus forms, the terms 'mull' 'mor' and 'moder' was introduced. Earthworms are generally rare or absent in mor humus form, but their occurrence in this form has been documented by Piearce (1972), Huhta and Koskenniemi (1975), Muys and Lust (1992). In the mull humus form earthworm populations are high. The activities of anecic and endogeic worms are major factors in removing organic materials from the soil surface and in maintaining the mining of materials into the mineral soil with subsequent stimulatory effects on soil microorganisms. The nature of the moder humus form results from the activities of epigeic earthworms (Rusek, 1985). In spite of mixing organic materials and incorporation of casts, they also create major changes in the surface organic layers which may have considerable effects on the communities of other litter inhabiting fauna and microorganisms (Parkinson and Mclean, 1998).

Many studies have been made to show that microbial species that are found by earthworm vary with circumstances (Doube and Brown, 1998). Stockdill (1982) showed that the lumbricid worm *Aporrectodae caliginosa* introduced in New Zealand pastures improve the soil qualities. Yeasts (1981) showed that the introduction of these worms into pasture soils lead to a 50% reduction in soil nematodes as well as change from bacterial to fungal - feeding species. Marinissen and Bok (1988) showed the effects of earthworms on the structure of polder soils which allow increased size, abundance and distribution of collembola in these soils. Impacts of earthworms on fungal and arthropod community structure has been studied by Parkinson (1988) and Scheu and Parkinson (1994 a,b). Earthworms also play an important role in the dispersal of beneficial and harmful soil microorganisms has been reviewed by Brown (1995) and Doube and Brown (1998). The majority of root diseases of agricultural crops are caused by soil-borne fungi. Doube *et al.* (1994 a,b) have shown that earthworm dispersed biological control agents can enhance control of these fungal diseases. Stephens *et al.* (1994 a,b) examimed the *Aporrectodae trapezoides* and *Aporrectodae rosea* which cause a significant reduction in the severity of the disease symptoms (root lesions) in both laboratory and field trials.

There is much evidence showing important role of earthworms in cycling of C, N and P in soils and modifying the soil microbial function (Hendrix *et al.*, 1987; Bostrom and Lofs- Holmin 1988; Parmelaee *et al.,* 1990, Lavelle and Martin 1992; Marinissen and de Ruiter, 1993). The size and composition of earthworm populations even within one locality vary with soil type and management practice (Lee 1985; Robinson *et al,* 1992). They also explained the earthworm dependent differences in the composition and abundance of the microbial communities which form the basis of the food web in soil. Gupta (1994) recognized that microbial populations are strongly influenced by soil pH and earthworms are sensitive to soil pH, being more abundant in neutral to slightly acid soil (Lofs-Holmin, 1986). Effects of the gut passage, the age of cast material and the type of ingested substrate on the mirobial community in *Lumbricus terrestris* faeces were studies by Tiunov and Scheu (2000). The abundance and organization of microathropod communities were correlated with the spatial distribution of the earthworm, *Polypheretima elongata.* The effects of the activities of the epigeic earthworm, *Dendrobaena octaedra*, on the oribatid community and microarthropod abundance in a pine forest was studied by Mclean and Parkinson (2000). Scullion and Malik (2000) described the earthworm activity affecting organic matter, aggregation and microbial activity in soils restored after opencast mining for coal. Because of these various activities, earthworms, if present in moderate biomass, are likely to have a substantial influence on the distribution, composition and activity of the microbial communities which are responsible for decomposition of organic residues and play a key role in regulating the biological processes which maintain soil health (Doube and Brown, 1998).

Sustainability

The world literature on earthworms in agricultural soils is huge. Much of the available information has been comprehensively reviewed by Edwards and Lofty (1977), Lee (1985) and Lavelle (1988). In southern Australia several studies have been made about the influence of the most common earthworm species of agricultural land on soil structure (Barley, 1959c; Hindell, *et al.*, 1994; Hirth *et al.*, 1994), nutrient availability (Barley and Jennings, 1959), burial of surface organic matter and lime (Barley, 1959b; Baker *et al.*, 1993, 1995), distribution of beneficial microorganisms (Doube *et al.*, 1994), reduction of root diseases (Stephens *et al.*, 1993, 1995) and plant yield and quality (Abbott and Parker, 1981; Temple - Smith *et al.*, 1993, Garnsey, 1994). Introduced lumbricid earthworms from Europe are also very common in agricultural habitats in other countries with temperate or mediterranean climates (Reinecke and Visser, 1980; Fender, 1985; Springett, 1992). The species that are most common in southern Australia agricultural soils are the introduced lumbricids *Aporrectodae caliginosa*, *A. trapezoides* and *A. rosea* which are also common in other regions of the world with similar climates. Like in other countries, soil acidity is a major environmental problem in high rainfall regions of Australia (Coventry, 1985; Chartres *et al.*, 1992). Lime is applied to the surface of soil to decrease acidity but it is generally slow to be incorporated into the root zones where it is needed (Helyar, 1991). Research in New Zealand (Stockdill and Cossens, 1966; Springett, 1983, 1985) has shown that some species of earthworms have the potential to bury lime and increase soil pH. Similarly, experiments have been conducted in south Australia to determine the species most likely to be useful in reducing soil acidity in this way. Influence of earthworms on the decomposition of plant material in soil of Scotland has also been analyzed (Cheshire and Griffiths, 1989).

Agricultural management practices such as tillage, crop rotation, stubble retention, drainage, irrigation, lime, fertilizer and slurry application, pesticide use and stocking rate can influence earthworm numbers and biomass (Lee, 1985; Lavelle *et al.*, 1989; Fraser, 1994). Rovira *et al.* (1987) demonstrated that the abundance of earthworms in a red brown earth soil in south Australia double by direct drilling of cereals compared with conventional cultivation. Fragaso *et al.* (1993) demonstrated a relationship between earthworm and soil organic matter in natural and managed ecosystems. Brussaard *et al.* (1993) worked on earthworm activity in relation to sustainability of agricultural systems. Brussaard (1994) also established interrelationships between biological activities, soil properties and soil management. Various experiments have showed the different abilities of some of the most common species to influence soil properties and plant production. (Baker, 1998). *Aporrectodae trapezoides* appears to be a much more beneficial species than *Aporrectodae rosea* in terms of promoting wheat grain yield and quality. Barley (1959a) showed that the numbers and mass of earthworms mainly of *A. rosea* and *A. trapezoides* increase with the addition of superphosphate to a pasture in south Australia. He argued that this leads to an increase in plant production and hence available food. Similarily, Fraser *et al.* (1994) found that earthworm numbers mostly of *A. caliginosa* and *L. rubellus* increase with superphosphate use and plant production in a New Zealand pasture. Effects of invasion of an aspen forest in the Canadian Rocky mountains by *Denrobaena octaedra* on nutrient mineralization, soil microflora and plant growth have also been investigated (Scheu and Parkinson, 1994b).

In India, the green revolution was promoted in 1960s to boost the food productions without foreseeing its ill effects. The recent realization about the need for ecological balance is essential for maintaining the sustenance of agricultural production. Farmers, scientists and policymakers are trying to find an alternative to chemical agriculture. In subtropical and tropical agricultural fields there is wider variation in the earthworm species than the species richness (Kale and Seenappa, 1997). Recently, Julka (2001) has nicely documented the distribution of earthworms in different agro-climatic regions of India. The functional role of earthworms in Indian agro-ecosystems has been also reviewed (Kale, 2001).

Land Bioreclamanation

Modern agricultural practices are resulting in an increasing impact on environment, causing a serious decline in the natural resources of arable lands. The biological effects of agricultural practices on a long term basis have received very little attention in the past. Management of natural resources on a sound ecological basis confronts us with problems of a biological origin (Christensen, 1988). The land bioreclamation deals with processes which determine the lives of organisms and is the subject to the action of living things.

Therefore, it is inevitable that the biological methods in land reclamation provides food possibilities for upgradation of conservation of soil fertility on a sustainable basis. Soil quality can be defined as the system of abiotic and biotic characteristics which secure the functioning of the soil environment. Due to close coupling of earthworms to processes and functions of ecosystems, land reclamation by earthworm is an exciting approach to improve the degraded lands in various parts of the globe.

The introduction of earthworms to lacking soils has resulted in significant increase in plant production in several countries. The inoculation of *A. caliginosa* to pastures in New Zealand, has initially increased pasture production by 72% and then by 25% in the longer term (Stockdill, 1982). The introduction of a mixture of *A. caliginosa, A. longa, Lumbricus rubellus* and *L. terrestris* in polder grasslands. The Netherlands has increased 10% productivity (Hoogerkamp *et al.,* 1983). Similarly, the introduction of *L. terrestris* and *A. caliginosa* to grassland on reclaimed peat bog in Ireland increased productivity 25-49% after two to three years (Curry and Boyl, 1987). Springett (1985) introduced *A. longa* to a pasture in New Zealand which already contained *A. caliginosa* and *L. rubellus* and recorded an increase in plant production of upto 27% in subsequent seasons. Edwards and Lofty (1980) inoculated *L. terrestris* and *A. Longa* into sterilized soil in England and recorded increased root growth and grain yield in direct drilled cereals. In contrast, James (1991) reported that the introduction of *A. caliginosa* and *Octolasion cyaneum* to tall grass prairie in the U.S.A. has a negative influence on soil properties through a reduction in the numbers of more useful native species (*Diplocardia* spp.).

The introduction of earthworms to new habitats have been reported by several workers in New Zealand and Europe (Stockdill 1982; Marinissen, 1991; Marinissen and van den Bosch, 1992; Stein *et al.,* 1992). There is a much potential to introduced beneficial species to sites lacking them either by further spread of species already present but patchily distributed within Australia, or the introduction of novel species or strains from climatically matched region overseas (Baker, 1998). The introduced lumbricid *A. longa* is currently restricted to Tasmania, but its potential distribution within Australia is predicted to be much larger. The introduction of *A. longa* into pastures in South Austria and Victoria suggest that substantial increases in pasture production may accrue from redistributing this species further and augmenting the earthworm fauna where deep-burrowing species are lacking (Baker, 1998).

The impacts of earthworms on soil processes and their potential utility in agriculture are currently of wide interest. Earthworms are popularly believed to reflect soil health. Lumbricid earthworms have been identical as bioindicator in different agroecosystems (Christensen, 1991). They often make up a large proportion of the biomass of the soil fauna and respond numerically to many agricultural management practices (Baker, 1998). Their abundance and biomass are also correlated with a range of edaphic variables. Earthworms have been suggested as potential indicators of the sustainability of agricultural practices that are useful for farmers (Oades and Walters, 1994). It is important that zoologists should recognize the earthworm species and realize their abilities to influence soil properties and plant production. Simple keys have been deviced for the common earthworm species in Australia that farmers might use (Baker and Barrett, 1994). With a great deal of care in sampling, earthworms might be combined with other physical and chemical indicators to monitor performance in local areas (Baker, 1998) In Indian condition *Polypheretima elongata* is suitable for soil reclaimation and produces more mucus with generation of available nutrients. It further helps in nutrient restoration and moisture conservation (Singh and Rai, 1998).

Nevative Aspects Into Profits

The earthworm potentials as a biological tool should be much better understood to make organic farming and sustainable development with the use of selected species of earthworms (Kale, 1998). Earthworms are specialized to live in decaying matter and can degrade it into fine particulate materials, high in available nutrients, with considerable potential as soil additives (Edwards, 1998). Earthworms are able to process sewage sludge and solids form waste water (Neuhauser *et al.*, 1988), brewery wastes (Butt, 1934), processed potato waste (Edwards, 1983), waste from the paper industries (Butt, 1993), wastes from supermarkets and restaurants (Edwards, 1995), wastes from the mushroom industry (Edwards, 1988), animal wastes from poultry, pigs, cattles sheep, goats, horses and rabbits (Edwards *et al.*, 1985, Edwards, 1988). Different laboratories have utilized earthworms to break down organic wastes which have been the main cause of organic pollution. The mechanisms responsible for stabilization of earthworm casts have also been investigated in laboratory (Haynes and Fraser, 1998).

Vermiculture has been receiving much importance in recent years in Canada, USA, Japan, Taiwan, Thailand and other countries (Govindan, 1998). Species which are identified as potentially useful species to break down organic wastes were *Eisenia fetida, Dendrobaena veneta* and *Lumbricus rubellus* from temperate areas and *Eudrilus eugeniae* and *Perionyx excavatus* from the tropics. The survival, growth, mortality and reproduction of these species were studied in detail in the laboratory, using a wide range of organic wastes (Edwards, 1998). Systems of growing earthworms range from very simple method involving low technology such as windows, waste heaps and boxes or through, moderately complex to completely automated continuous flow reactors (Jensen, 1993, Edwards, 1995). The growth of *Eisenia fetida, Eudrilus eugeniae, Perionyx excavatus* and *Perionyx hawayana* in sewage sludge has been studied (Neuhauser *et al.*, 1988). Worms are also bred commercially on a large scale in organic waste for fish bait. The species most commonly bred in this way is *Eisenia fetida* (Savigny). Work in the United States, at the State University of New York has shown that earthworms can breakdown activated sludges very effectively (Hartenstein *et al.*, 1979). The key to maximum productivity is to maintain aerobicity and optimum moisture (80-90%) and temeperature (15-20%) and pH between 5 and 9 (Edwards, 1988) Degradation of organic waste using a number of species viz., *Eisenia fetida, Aporrectodea caliginosa, Eudrilus eugeniae, Perionyx arboricola* etc has been reported from different parts of India other than world's region (Goswami and Kalita 2000). Frederickson *et al.* (1997) have combined vermiculture with traditional green waste composting systems.

The use of earthworms in sludge management has been termed vermicompost or vermistabilization (Neuhauser *et al.*, 1988, Loehr *et al.*, 1984). Earthworms compost (vermicomposts) can be produced from almost all kinds of organic wastes with suitable preprocessing and controlled processing conditions. Fosgate and Babb (1972) grew earthworms in cattle waste and reported that the vermicomposting produce was equal to greenhouse potting mixes for production of flowering plants. Julka and Mukherjee (1986) demonstrated the potentiality of an earthworm in decomposition of organic matter of sterilized soil. Kale and Bano (1986) made field trials of vermicompost and found it as effective as chemical fertilizers. Reddy (1988) reported increased growth of *Vinca rosea* and *Oryza sativa* after addition of cast material from *Pheretima alexendri*. The vermicomposts had higher levels of available nutrients than the waste from which they were formed (Buchanam *et al.,* 1988). Edwards (1988) reported that vermicomposts had high levels of available nitrogen. Jambhekar (1990) found that the application of vermicomposts increased the available N, P and K content in soil. The influence of vermicompost application on the available macronutrients and selected microbial population in a paddy field was studied by Kale *et al.* (1992). Hapse *et al.* (1993) observed that application of vermicompost @ 5 mg/ha significantly increased total N, available N.P.K. and organic carbon and decreased pH over control. They grow plants extremely well, and they can also be used as structural additive for poorer soils to provide nutrients and minimize erosion (Edwards, 1998). The earthworm tissues contain about 72 percent of proteins (Senapati and Dash, 1984). These proteins have arginine 4 times that in blood meal and seven times in beef liver. Earthworm protein is also about 2.5 times richer in tyrosin as compared to liver protein. The earthworms have been found to act as a good poultry food (Hardwood, 1976; Yashida and Hoshi 1978; Mekada *et al.,* 1979, Taboga, 1980).

The earthworms are included in the diets of pigs (Harwood and Sabine, 1978) and of mice and rats (Mclnroy, 1971; Schulz and Graff, 1977). The earthworms are also used as fish feed (Arunachalam and Palanichamy, 1984). The casts of earthworms have been found to contain plant growth promoters (Krishnamoorthy and Vajaranabhaiah, 1984). Vermicomposting biotechnology is must for our country in order to minimize environmental pollution and obtain good organic fertilizers. In spite of giving high quality of protein for both live stock and aquaculture, it is also an income generating technology for improving the socio-economic conditions of the farmers and weaker sections of the society (Madan and Sharma, 1986). Guerrero (1983) reported that *Tilapia* fish grow better on diets containing earthworm protein supplements from *Perionyx excavatus* than those provided with other fish meal supplements. Velasquez *et al.* (1991) reported that feed prepared from *Eisenia fetida* produces satisfactory growth of rainbow trout with a significant increase in lipid content.

Vermitek Systems of Canada and Oregon Soil Corporation of U.S.A. are the two organizations actively engaged in developing an automated system or hightech system of vermicomposting (Jensen, 1993). Earthworms belonging to epigeic group possesses higher composting power (Bhole, 1992 and Senapati, 1994). Goswami and Kalita (2000) showed the efficiency of some indigenous earthworms species of Assam and their role in vermitechnology According to them *Amynthas diffringens* is the best species for decomposing organic wastes. In India the vermitechnology process has been evaluated for treatment of agro, sugar, food processing and other wastes (Kale, 2000). Recycling of wastes through vermitech reduces the problem of non-utilization of agrowaste. Nutrients present in vermicompost are readily available and the increase in earthworm populations on a application of vermicompost and mulching lead to the easy transfer of nutrients to the plants providing synchrony in ecosystems (Ismail, 2000).

Toxicological Assessments

The modern society has become increasely dependent on the use of chemicals that are harmful to the environment and agriculture. Since then there is an awakening and gradual increase in information about the side effects of chemicals. Concerns about soil fertility, risks of chemical leaching into drinking water, contamination of soil and detrimental effects of contaminates on the living environment have resulted in a strong and growing interest in soil organisms among environmental biologists and legislators. Many toxic materials accumulate along food webs and the detrivore - decomposers levels are frequently the first to be effected since the organic matter and soil are the ultimate sink for most contaminate. Legislation in many countries has recently focused the attention of workers on the need for sensitive organisms from the soil environment for use in research and environmental monitoring (Reinecke and Reinecke, 1998). No doubt earthworms may play an important role in ecotoxiological evaluation. In spite of doubts being *Eisenia fetida* as typical earthworm in toxiological studies, the position of this species in acute toxicity tests seems to be secured.

Contaminants exert their effects at all levels of biological organization but research pertaining to environmental toxicology has concentrated on understanding effects at individual and lower levels. Much efforts has been put into earthworms test development (Eijsackers, 1998). The Organisation for Economic Coordination and Development (OECD) and the International Standards Organisation (ISO) have carried out a number of tests and standardization studies. Not with standing all these efforts, in a recent studies for OECD on the selection of a set of laboratory ecotoxicity test, the earthworms tests secured 6th position for the OECD tests, 13 for the test of Environmental Protection Agency (EPA), and 18 for the ISO tests Standardization of earthworms for toxicity tests has also been done by the International Organization for Biological Control (IOBC) and the National Institute for Human Health and the Environment (RIVM-NL). If more ecologically relevant earthworm tests should be accepted, a lot of work has to be done. Only in this way an ecological sound risk assessments of environmental contaminants and potential contaminating activities and situations be achieved (Eijsackers, 1998).

Stenersen *et al.* (1973) studied the toxicity of carbofuran to the earthworm *Lumbricus terrestris*. The laboratory toxicity of several insecticides on four different species of earthworms was investigated by Sterersen (1979). *Eisenia fetida* was the most tolerant species to the pesticide tested and the aldicarb was the most toxic pesticide to this species. Lofes (1980) reported that trichloroacetic acid and benomyl retarde the growth and cause heavy mortality of *Allolobophora chlorotica, Lumbricus terrestris* and *A. caliginosa*. Cathey (1982) evaluated comparative toxicity of DDE, aldrin, endrin, parathion and carbaryl to the earthworm L. *terrestris*. Banhawy *et al.* (1984a,b) reported that dimethoate and curacron induce marked morphological signs of poisoning and permanent histological effects in the intestine and nerve cord of *A. calinginosa*. Reddy *et al.* (1984) showed oozing of coelomic fluid from the body, swelling of clitellum and posterior segment and decrease in carbohydrate and glycogen contents in *L. mauritii* in response to fenitrothion exposure. Neuhauser *et al.* (1985) described the toxicity of phenols, amines, substituted aromatics, halogenated aliphatics, ether, polycyclic aromatic hydrocarbon and pthalates to *E. fetida*. Bharathi and Rao (1986) noted the effect of monochrotophas and dichlorovas to *L. mauritii*.

Neuhauser *et al.* (1986) evaluted comparative toxicity of a number of chemical compounds to *A. tuberculata, E. fetida, Eudrilus eugeniae* and *Perionyx excavatus*. Kula and Kokta (1992) monitered the side effects of some pesticides on earthworms under laboratory and field conditions. Maurya and Chattoraj (1994) studied toxicological symptoms of various doses of carbaryl, aldrin, rogor and lindane on *Perionyx sansibaricus* and *Metaphire*. The toxicological symptoms observed in response to pesticides were body coiling, swelling of clitellar or other body region, reddening of dorsal blood vessel, oozing of coelomic fluids, appearance of ring like constriction and damage of dermal skin. They noted that the severance of contractions of body segments occurred only in *Perionyx sansibaricus*. However, this particular phenomenon was absent in *M. posthuma*. Mohamed *et al.* (1995) studied the impact of methomyl, lebaycid, paraquat, glyphosate and remiltine on the survival and body mass of the earthworm *A. caliginosa*.

Bennour *et al.* (1997) evaluated the effects of pomex, formothion, fenvalerate,. glyphosate and ridomil on the survival, body mass and transpiration of *Aporrectodea caliginosa*. They found pomex and fenvalerate as a highly toxic insecticides to the earthworm. Approach is being made to estimate the bioconcentration of organic chemical in earthworms (Jager, 1998). The effects of benomyl application on the reproductive rate of *A. rosea* and *A. longa* have also been studied (Holmstrup *et al.*, 1998). The pesticide effects on populations of some earthworm species have been evaluated (Axelsen and Holmstrup, 1998; Holmstrup, 2000).

In the last decades, some important biochemical studies on earthworms have been reported (Albro *et al.*, 1992; Holmstrup and *et al.*, 1999; Flegel and Schrader, 2000; Petersen and Holmstrup, 2000). Goven *et al.* (1994) studied lysozyme activity in *L. terrestris* coelomic fluid and coelmocytes for assaying immunotoxicity of xenobiotics. Earthworms immuno assays have been developed for evaluating biological effects of exposure to hazardous materials (Fitzpatrick *et al,* 1990). Drewes and Lingamneni (1992) analysed the use of earthworm in econeurotoxicity. Edwards (1992) discussed the advantages and limitations of ecotoxicological tests on earthworm. Reinecke and Reinecke (1998) suggested the use of earthworms in toxicological evaluation as a new approach in risk assessment.

Summary

Soil bioresources have been recognized as the foundation for sustainable livelihood, food security and environmental safety. The importance of earthworms at this stage cannot be ignored because they have enormous potentials to improve soil condition and clean environment on a sustainable basis. In the recent years the print and electronic media have popularized the importance of earthworms worldwide. The publication of popular articles in newspapers and magazine and scientific deliberations have attracted the attention of policy makers and administraters and ultimately scientists and environmental engineers are encouraged to work on earthworm bioresources and vermicomposting for a better and sustainable soil environment. Rapid and unsafe growth of industries in the past century has considerably deteriorated the biosphere. Decline in natural resources and increase in hazardous chemicals in environment is estimated to cause a substantial loss of Gross Domestic Production (GDP) of various countries. As a result the agricultural land is losing its productive top soil 20-40% times faster than the soil naturally can reform in thousands of years. The fossil – energy based agricultural inputs have drawn down energy budget and questioned our future food security and generational equity. It may be effectively minimized by popularizing the use of biofertilizers and biopesticides and utilizing the potentials of earthworm bioresources. The living planet earth is very rich in earthworm resources and its climate is suitable for propagation of vermicomposting species. There is a good possibility that some of the earthworm species may be identified suitable for development of earthworm bioindustry. Earthworm have also been recognized to reduce atmospheric level of carbon dioxide. The work on earthworm resources will integrate technology with rural development for upliftment of regional economies and preserving natural environment.

REFERENCES

Abbott, I. and Parker, C.A. (1981) Interactions between earthworms and their soil environment. *Soil Biol. Biochem.*, 13: 191-197.

Alban, D.H. and Berry, E.C. (1994) Effects of earthworm invasion on morphology, carbon and nitrogen of a forest soil. *Appl. Soil Ecol.*, 1: 243- 249.

Albro, P.W., Schroeder, J.L. and Corbett, J.T. (1992) Lipids of the earthworm *Lumbricus terrestris. Lipids,* 27: 136-143.

Aritajat, U., Madge, D.S. and Gooderham, P.T. (1977) The effect of compaction of agricultural soils on fauna. I. Field investigations. *Pedobiologia*, 17: 262-282.

Arldt, T. (1908) Die Ausbreitung der terricolen Oligochaeten in Laufe der erdgeschichtlichen Enwicklung des Erdreliefs. *Zool.Jb. (Syst.),* 26: 285-318.

Arunachalam, S. and Palanichamy, S. (1984) Earthworms as feed for the catfish *Mystus vittatus*. A paper presented to *Natl. Sem.Org. Waste Utiliz. Vermicomp.* School of Life Sciences, Sambalpur University, Jyoti Vihar, Orrisa. 4 p.

Axelsen, J.A. and Holmstrup, M. (1998) Simulation of pesticide effects on populations of 3 species of earthworms. In: *Advances in Earthworms Ecotoxicology. Proc. from 2nd Int Workshop on Earthworm Ecotoxicology* 2-5 April 1997. Amsterdam, The Netherlands Pensacole. FL. *Soc. Environ. Toxicol. Chem.*, pp. 281-29

Bahl, K.N. (1950) *The Indian Zoological Memoirs.* I. *Pheretima.* 4th edition. Lucknow Publishing House, Lucknow, India. 84 p.

Baker, G.H. (1998) The ecology, management and benefits of earthworms in agricultural soils with particular reference to southern Australia. In: *Earthworm Ecology* (C.A. Edwards, ed.), pp. 230-257. CRC Press, The Netherlands.

Baker, G.H. and Barrett, V.J. (1994) Earthworm Identifier. CSIRO, Melbourne, Australia. pp. 1-12.

Baker, G.H., Barrett,V.J., Carter, P.J., Buckerfield, J.C.,Williams, P.M.L. and Kilpin, G.P. (1995) Abundance of earthworms in soils used for cereal production in south- eastern Australia and their role in reducing soil acidity. In: *Plant Soil Interactions at Low pH* (R.A. Date, N.T. Grundon, G.E. Rayment and M.E. Probert, eds.), pp. 213-213. Kluwer Academic Publishers, Dordrecht.

Baker, G.H., Williams, P.M.L., Barrett, V.J. and Carter, P.J. (1993) Burial of surface-applied lime by earthworms in south Australian pastures. *Proc. 17th Int. Grassland Cong.*, 1: 939-941.

Banhawy, M.A., Ganzuri, M.A. and E1- Akkad, M.M. (1984a) Morphological and histological studies on the effects of insecticides on the earthworms, *A. caliginosa. Bull. Zool. Soc. Effect*, 9: 27-31.

Banhawy, M.A., Ganzuri, M.A. and E1- Akkad, M.M. (1984b) Effects of insecticides on the histological structure of the intestinal wall of the earthworms, *Allolobophora caliginosa*. *Ann. Zool.* (India), 23: 16-31.

Barley, K.P. (1959a) The influence of earthworms on soil fertility. I. Earthworm populations found in agricultural land near Adelaide. *Aust. J. Agric. Res.*, 10: 171-178.

Barley, K.P. (1959b) The influence of earthworms on soil fertility. II. Consumption of soil and organic matter by earthworm *Allolobophora caliginosa*. *Aust. J. Agric. Res.*, 10: 179-185.

Barley, K.P. (1959c) Earthworms and soil fertility. IV. The influence of earthworms on the physical properties of a red-brown earth. *Aust. J. Agric. Res.*, 10: 371-376.

Barley, K.P. and Jennings, A.C. (1959) Earthworms and soil fertility. III. The influence of earthworms on the availability of nitrogen. *Aust. J. Agric. Res.*, 10: 364-370.

Beare, M.H., Coleman, D.C., Crossley, D.A. Jr., Hendrix, P.F. and Odum, E.P. (1994) A hierarchical approach to evaluating the significance of soil biodiversity to biogeochemical cycling. *Plant and Soil,* 170: 5-22.

Bhadauria, T. and Ramakrishnan, P.S. (1989) Earthworm population dynamics and contribution to nutrient cycling during cropping and fallow phases of shifting agriculture (jhum) in north-east India. *J. Appl. Ecol.*, 26: 505-520.

Bharathi, Ch. and Subha Roa, B. V.S.S.R. (1986) Toxic effects of two organophosphate insecticides, Monocrotophos and Dichlorvos to common earthworm *Lampito mauritii* (Kinberg). *Water Air Soil Pollut.*, 28: 128-130.

Bhole, R.J. (1992) Natural farming and vermiculture biotechnology. In: *Proc. Natl. Sem. Org. Farm. held at College of Agril., Pune.* pp. 28-29.

Blair, J.M., Parmelee, R.W., Allen, M.F., McCartney, D.A. and Stinner, B.R. (1997) Changes in soil N pools in response to earthworm population manipulations in agroecosystems with different N sources. *Soil Biol. Biochem.*, 29: 361-367.

Block, W. and Banage, W.B. (1968) Population density and biomass of earthworms in some Uganda soils. *Rev. Ecol. Biol. Sol.*, 5: 515-521.

Bornebusch, C.H. (1930) The fauna of the forest soil. *Forstl. Forsogsv. Dan.*, 11: 1-224.

Bostrom, U. and Lofs-Holmin, A. (1988) Earthworm population dynamics and flows of carbon and nitrogen through *Aporrectodea caliginosa* (Lumbricidea) in four cropping systems. In: *Ecology of Earthworms in Arable Land: Population Dynamics and Activity in Four Cropping Systems*.Institutionen for ekologi och miljovard, Report No. 34. Swedish University of Agricultural Sciences, Uppsala .

Bouche, M.B. (1969) La biogeographie des lumbricides de France, son interet et ses ambiguites. Cas d'*Allolobophora cupulifera* Tetry, *A. icterica* (Sav.), de *Lumbricus friendi* Cognetti et de *Lumbricus herculeus* (Sav.). *Pedobiologia*, 9: 87-92.

Bouche, M.B. (1972) Lombriciens de France: Ecologie et systematique. *Ann. Zool. Ecol. Anim.*, Numero special, 72: 1-671.

Bouche, M.B. (1975) Fonctions des lombriciens. III. Premieres estimations quantitative des stations francaises du P.B.I. *Rev. Ecol. Biol. Sol.*, 12: 25-44.

Bouche, M.B. (1977) Strategies lombricienns. *Ecol. Bull.*(Stockholm), 25: 122-132.

Bouche, M.B. (1981) Development et lombriciences: Biostimulation des soils et bio-indicator. In: *Compte rendu des Journees Science ecologigue et development.*, pp. 281-295. AFIE, Grenoble.

Bouche, M.B. (1983) The establishment of earthworm communities. In: *Earthworm Ecology-from Darwin to Vermiculture* (J.E. Satchell, ed.), pp. 431-447. Chapman and Hall, London, England.

Bouche, M.B., Al-Addan, F., Cortez, J., Hameed, R., Heidet, J.C., Ferriere, G., Mazaud, D. and Samih, M. (1997) Role of earthworm in N cycle: A falsifiable assessment. *Soil Biol. Biochem.*, 29: 375-380.

Boyd, J.M. (1958) The ecology of earthworms in cattle-grazed machair in Tiree, Argyll. *J. Anim. Ecol.*, 27:147-157.

Brown, G.G. (1995) How do earthworms affect microfloral and faunal community diversity? *Plant and soil*, 170: 209-231.

Bruning, J.L. and Kintz, B.L. (1977) Analysis of variance. In: *Computational Handbook of Statistics* (Second Edition), pp. 18-406. Scott, Foreman and Company, Gleen view, Ilionois, USA.

Brussaard, L. (1994) Interrelationships between biological activites, soil properties and soil management. In: *Soil Resilience and Sustainable Land Use* (D. J. Greenland and I. Szabolcs, eds.), pp. 309-329. CAB International, The Netherlands.

Brussaard, L., Hauser, S. and Tian, G. (1993) Soil faunal activity in relation to the sustainablility of agricultural systems in the humid tropics. In: *Soil Organic Matter Dynamics and Sustainability of Tropical Agriculture* (K. Mulongoy and R. Merckx, eds.), pp. 241-245. A Wiley-Sayce Co. Publication, New York.

Buchanam, M.A., Rusell, E. and Block, S.D. (1988) Chemical characterization and nitrogen mineralization potentials of vermicomposte derived from differing organic wastes. In: *Earthworms in Environmental and Waste Management* (C.A., Edwards and E.F. Neuhauser, eds.), pp. 231-240. SPB Academic Publishing, The Netherlands.

Butt,K.R. (1993) Utilization of solid paper mill sludge and spent brewery yeast as a feed for soil-dwelling earthworms. *Bioresour. Tech.,* 44: 105-107.

Cathey, B. (1982) Comparative toxicities of insecticides to the earthworm *Lumbricus terrestris.Agric. Environ.*, 7:73-81.

Chappell, H.G., Ainsworth, J.F., Cameron, R.A.D. and Redfern, M. (1971) The effect trampling on a chalk grassland ecosystem. *J. Appl. Ecol.,* 8: 869-882.

Chartres, C.J., Helyar, K.R., Fitzpatrick, R.W. and Williams, J. (1992) Land degradation as a result of European settlement of Australia and its influence on soil properties. In: *Australia's Renewable Resources: Sustainability and Global Change.* Bureau of Rural resources *Proceedings* No. 14. (R.M. Gifford and M.M. Barson, eds.), pp. 3-33. Government Printing Service, Canberra.

Chauhan, T.P.S. (1980) Seasonal changes in the activities of some tropical earthworms. *Comp. Physiol. Ecol.*, 5: 288-298.

Cheshire, M.V. and Griffiths, B.S. (1989) The Influence of earthworms and cranefly larvae on the decomposition of uniformly 14 C levelled plant material in soil. *J. Soil Sci.,* 40: 117-123.

Christensen, O. (1988) The direct effects of earthworms on nitrogen turnover in cultivated soils. *Ecol. Bull. (Copenhagen)*, 39: 41-44.

Christensen, O. (1991) Lumbricid earthworms as bioindicators relative to soil factors in different agroecosystems. In: *Advances in Management and Conservation of Soil Fauna* (G.K. Veeresh, D. Rajagopal and C.A. Viraktamath, eds.), pp. 839-849. Oxford and IBH Publishing Co. Pvt. Ltd., New Delhi.

Conventry, D.R. (1985) Changes in agricultural systems on acid soils in southern Australia. *Proc. 3rd Aust. Agro. Conf.,* pp. 126-145.

Curry, J.P. and Boyle, K.E. (1987) Growth rates, establishment and effects on herbage yield of introduced earthworms in grassland on reclaimed cutover peat. *Biol. Fertil. Soil*, 3: 95-98.

Darwin, C. (1881) *The Formation of Vegetables Mould through the Action of Worms with Observations on their Habits.* John Murray, London, 326 p.

Dash, M.C. and Patra, V.C. (1977) Density biomass and energy budget of a tropical earthworm population from a grassland site in Orissa, India. *Rev. Ecol. Biol. Sol.,* 14: 461-471.

Dash, M.C. and Patra, V.C. (1979) Wormcast production and nitrogen contribution to soil by a tropical earthworm population from a grassland site in Orissa, India. *Rev. Ecol. Biol. Sol.,* 16: 79-83.

Dash, M.C. and Senapati, B.K. (1980) Cocoon morphology, hatching and emergence pattern in tropical earthworms. *Pedobiologia,* 20: 316–324.

Devliegher, W. and Verstraete, W. (1997) The effect of *Lumbricus terrestris* on soil in relation to plant growth: Effects of nutrient-enrichment processes (NEP) and gut- associated processes (GAP). *Soil Biol. Biochem.*, 29: 341-346.

Doube, B.M. and Brown, G.G. (1998) Life in a complex community: Functional interactions between earthworms, organic matter, microorganisms and plants. In: *Earthworm Ecology* (C.A. Edwards, ed.), pp. 179-211. CRC Press, The Netherlands.

Doube, B.M., Stephens, P.M., Davoren, C.W. and Ryder, M.H. (1994a) Interactions between earthworms, beneficial soil microorganisms and root pathogens. *Appl. Soil Ecol.*, 1: 3-10.

Doube, B.M., Stephens, P.M., Davoren, C.W. and Ryder, M.H. (1994b) Earthworms and the introduction and management of beneficial soil microorganisms. In: *Soil Biota: Management in Sustainable Farming Systems* (C.E. Pankhurst, B.M. Doube, V.V.S.R. Gupta and P.R.Grace, eds.), pp. 32-41. CSIRO, East Melbourne, Australia.

Drewes, C.D. and Lingamneni, A. (1992) Use of earthworms in eco-neurotoxicity: effects of carbofuran in *Lumbricus terrestris.* In: *Ecotoxicology of Earthworms* (P.W. Greig-Smith *et al.*, eds.), pp. 63-72. Intercept Ltd., Hants, U.K.

Dzangaliev, A.D. and Belousova, N.K. (1969) Earthworm populations in irrigated orchards under various soil treatments. *Pedobiologia*, 9: 103-105.

Edwards, C.A. (1983) Earthworm ecology in cultivated soils. In: *Earthworm Ecology-from Darwin to Vermiculture* (J.E. Satchell, ed.), pp. 123-127. Chapman and Hall, London, England.

Edwards, C.A. (1988) Breakdown of animal, vegetable and industrial organic wastes by earthworm. In: *Earthworms in Waste and Environmental Management* (C.A. Edwards and E.F. Neuhauser, eds.), pp. 21-31. SPB, The Hague.

Edwards, C.A. (1992) Testing the effects of chemicals on earthworms: the advantages and limitations of field tests. In: *Ecotoxicology of Earthworms* (P.W. Greig- Smith *et al.*, eds.), pp. 75-84. Intercept Ltd., Hants, U.K.

Edwards, C.A. (1995) Commercial and environmental potential of vermicomposting: A historical overview. *Biocycle*, pp. 62-63.

Edwards, C.A. (1998) The use of earthworms in the breakdown and management of organic wastes. In: *Earthworm Ecology* (C.A. Edwards, ed.), pp. 327-354. CRC Press, The Netherlands.

Edwards, C.A. and Heath, G.W. (1963) The role of soil animals in breakdown of leaf material. In: *Soil Organisms* (J. Doeksen and Van der Drift, eds.), pp. 76-80. North Holland Publishing Co., Amsterdam, The Netherlands.

Edwards, C.A. and Lofty, J.R. (1977) *Biology of earthworms.* Chapman and Hall, New York. 333 p.

Edwards, C.A. and Lofty, J.R. (1980) Effects of earthworm inoculation upon the root growth of direct drilled cereals. *J. Appl. Ecol.,* 17: 533-543.

Edwards, C.A., Burrows, I., Fletcher, K.E. and Jones, B.A. (1985) The use of earthworms for composting farm wastes. In: *Composting of Agricultural and Other Wastes* (J.K.R. Gasser, ed.), pp. 229-242. Elsevier, Amsterdam, The Netherlands.

Edwards, C.A., Thompson, A.R. and Lofty, J.R. (1967) Changes in soil invertebrates populations caused by some organophosphate insecticides. *Proc. 4th Brit. Insecticide Fungicide Conf.*, pp. 48-55.

Edwards, W.M. and Shipitalo, M.J. (1998) Consequences of earthworms in agricultural soils. Aggregation and porosity. In: *Earthworm Ecology* (C.A. Edwards, ed.), pp. 147-161. CRC Press, The Netherlands.

Eijasackers, H. (1998) Earthworms in environment research: Still a promising tool. In: *Earthworm Ecology* (C.A. Edwards, ed.), pp. 295-323. CRC Press, The Netherlands.

El- Duweini, A.K. and Ghabbour, S.I. (1965) Population density and biomass of earthworms in different types of Egyptian soils. *J. Appl. Ecol.,* 2: 271-287.

Eswaran, H. (1994) Soil resilience and sustainable land management in the context of AGENDA 21. In: *Soil Resilience and Sustainable Land Use* (D.J. Greenland and I. Szaboles, eds.), pp. 36-49. CAB International, Wallingford, U.K.

Evans, A.C. and Guild, W. J. Mc. L. (1947) Studies on the relationships between earthworms and soil fertility. I. Biological studies in the field. *Ann. Appl . Biol.,* 34: 307-330.

Evans, A.C. and Guild, W.J. Mc. L. (1948) Studies on the relationships between earthworms and soil fertility. IV. On the life cycles of some British Lumbricidae. *Ann. Appl. Biol.,* 35: 471-484.

Fender, W.M. (1985) Earthworms of the western United States. Part I. Lumbricidae. *Megadrilogica,* 4: 93-129.

Fitzpatrick, L.C., Goven. A.J., Venables, B.J. and Cooper, E.L. (1990) Earthworms immunoassays for evaluating biological effects of exposure to hazardous materials. In: *In Situ Evaluaition of Biological Hazards of Environmental Pollutants* (S.S. Sandhu, ed.), pp. 119-129. Plenum Press, New York.

Flegel, M. and Schrader, S. (2000) Importance of food quality on selected enzyme activities on earthworm casts (*Dendrobaena octaedra*. Lumbricidae) *Soil Biol. Biochem.*, 32: 1191-1196.

Fosgate, O.T. and Babb, M.R. (1972) Biodegradation of animal waste by *Lumbricus terrestris. J. Dairy Sci.*, 55: 870-872.

Fragoso, C. and Lavelle, P. (1992) Earthworm communities of tropical rain forests. *Soil Biol. Biochem.*, 24: 1397-1408.

Fragoso, C., Barois, I., Gonzalez, C., Arteaga, C. and Patron, J.C. (1993) Relationship between earthworms and soil organic matter levels in natural and managed ecosystems in the Mexican tropics. In: *Soil Organic Matter Dynamics and Sustainability of Tropical Agriculture* (K. Mulongoy and R. Merckx, eds.), pp. 231-239. A Wiley-Sayce Co. Publication, New York.

Fraser, P.M. (1994) The impact of soil and crop management practices on soil macrofauna. In: *Soil Biota: Management in Sustainable Farming Systems* (C.E. Pankhurst, B.M. Doube, V.V.S.R. Gupta and P.R. Grace, eds.), pp. 125-132. CSIRO, Melbourne, Australia.

Fraser, P.M., Haynes, R.J. and Williams, P.H. (1994) Effect of pasture improvement and intensive cultivation on size of microbial biomass, enzyme activities and composition and size of earthworm populations. *Biol.Fertil. Soil,* 17: 185-190.

Frederickson, J., Butt, K.R., Morris, R.M. and Daniel, C.(1997) Combining vermiculture with traditional green waste composting systems. *Soil Biol. Biochem.*, 29: 725-730.

Garnsey, R.B. (1994) Increasing earthworms population and pasture production in the midlands of Tasmania through management and the introduction of the earthworms *Aporrectodea longa*. In: *Soil Biota: Management in Sustainable Farming Systems. Poster paper* (C.E. Penkhurst, ed.), pp. 27-30. CSIRO, Melbourne, Australia.

Gates, G.E. (1945c) Another species of *Pheretima* from India. *Sci. Cult.* Calcutta, 10: 403.

Gates, G.E. (1972) Burmese earthworms. An introduction to the systematics and biology of megadrile oligochaetes with special reference to southeast Asia. *Trans. Am. Phil. Soc.,* 62: 1-326.

Gerard, B.M. (1967) Factors affecting earthworms in pasture. *J. Anim. Ecol.,* 36: 235-252.

Ghabbour, S.I. and Shakir, S.H. (1982) Population density and biomass of earthworms in agro- ecosystems of the Meriut coastal desert region in Egypt. *Pedobiologia*, 23: 189-198.

Gilot, C. (1997) Effects of a tropical geophagous earthworm, *M. anomala* (Megascolecidae), on soil characteristics and production of a yam crop in Cote d' Ivoire. *Soil Biol. Biochem.*, 29: 353-359.

Goswami, B. and Kalita, M.C. (2000) Efficiency of some indigenous earthworms species of Assam and its characterization through vermitechnology. *Indian J. Environ. Ecoplan.*, 3: 351-354.

Goven, A.J., Chen, S.C., Fitzpatrick, L.C. and Venables, B.J. (1994) Lysozyme activity in earthworm (*Lumbricus terrestris*) coelomic fluid and coelomocytes: enzyme assay for immunotoxicity of xenobiotics. *Environ. Toxicol. Chem.*, 13: 607-613.

Govindan, V.S. (1998) Vermiculture and vermicomposting. In: *Ecotechnology for Pollution Control and Environmental Management* (R.K. Trivedy and A. Kumar, eds.), pp. 49-58. Enviro Media, Karad, India.

Graff, O. (1953) Investigations in soil zoology with special reference to the terricole Oligochaeta. *Z. Pfi Ernahr.Dung.*, 61: 72 –77.

Guerrero, R.D. (1983) The culture and use of *Perionyx excavatus* as a protein resource in the Philippines. In: *Earthworms Ecology- from Darwin to Vermiculture* (J.E. Satchell, ed.), pp. 309-313. Chapman and Hall, London, England.

Guild, W.J.Mc.L. (1948) Studies on the relationships between earthworms and soil fertility. III. Effect of soil type on populations. *Ann. Appl. Biol.,* 35: 181-192.

Guild, W.J.Mc.L. (1951) The distribution and population density of earthworms (Lumbricidae) in Scottish pasture fields. *J. Anim. Ecol.,* 20: 88-97.

Guild, W.J.Mc.L. (1952) Variation in earthworm numbers within field populations. *J. Anim. Ecol.*, 21: 169.

Guild, W.J.Mc.L. (1955) Earthworms and soil structure. In: *Soil Zoology* (D.k.Mc.E.Kevan, ed.), pp. 83-98. Butterworths, London, England.

Gupta,V.V.S.R. (1994) The impact of soil and crop management practices on the dynamics of soil microfauna and mesofauna. In: *Soil Biota: Management in Sustainable Farming Systems* (C.E. Pankhurst, B.M. Doube, V.V.S.R. Gupta and P.R. Grace, eds.), pp. 107-124. CSIRO, East Melbourne, Australia.

Hahta, V., Karppinen, E., Nurminen, M. and Valpas, A. (1967) Effect of silvicultural practices upon arthropod, annelid, and nematode populations in coniferous forest soils. *Ann. Zool. Fenn.*, 4: 87-143.

Haimi, J. and Boucelhom, M. (1991) Influence of a litter feeding earthworms, ***Lumbricus rubellus***, on soil processes in a simulated coniferous forest floor. ***Pedobiologia***, 35: 247-256.

Haimi, J. and Huhta, V. (1990) Effects of earthworms on decomposition processes in raw humus soil: a microcosm study. *Biol. Fertil. Soil*, 10: 178-181.

Hapse, P.G., Murkute, S.B. and Zende, N.A. (1993) Effect of vermicompost on sugarcane yield and sugar recovery "*10th Annual State Level Sugarcane Development Workshop on Low Cost Technology for Cane and Sugar Production*" Organized by V.S.F. Manjari (B.K.) on 9-10th July, 1993.

Hardwood, M. (1976) Recovery of protein from poultry waste by earthworms. *Proc. Ist Aust. Poult. Stockfed. Conv*. Melbourne, pp. 138-143.

Hardwood, M. and Sabine, J.R. (1978) The nutritive value of worm meal. In: *Proc. of 1st Aust. Poult. Stockfed. Conv*. Sydney, pp. 164-171.

Hartenstein, R., Neuhauser, E.F. and Kaplan, D. (1979) Reproductive potential of the earthworm *Eisenia foetida*. ***Oecologia***, 43: 329-340.

Haynes, R.J. and Fraser, P.M. (1998) A comparison of aggregate stability and biological activity in earthworm casts and uningested soil as affected by amendment with wheat or lucerne straw. *Eur. J. Soil Sci.*, 49: 629-636.

Helyar, K.R. (1991) The management of acid soils. In: *Plant-Soil Interactions at low pH*. (R.J. Wrigth, V.C. Baligar and R.P. Murrmann, eds.), pp. 365-382. Kluwer Academic Publishers, Dordrecht.

Hendrix, P.F., Crossley, D.A. Jr., Coleman, D.C., Parmelee, R.W. and Beare, M.H. (1987) Carbon dynamics in soil microbes and fauna in conventional and no-tillage agroecosystems. *INTECOL Bulletin*, 15: 59-63.

Hindell, R.P., Mckenzie, B.M. and Tisdall, J.M. (1994) Destabilisation of soil aggregates by geophagous earthworms, a comparison between earthworm and artificial casts. In: *Soil Biota:Management in Sustainable Farming Systems. Poster Papers* (C.E. Pankhurst, ed.), pp. 131-132. CSIRO, Melbourne, Australia.

Hirth, J.R., Mckenzie, B.M. and Tisdall, J.M. (1994) Some physical characteristics of earthworms casts in burrows. In: *Soil Biota: Management in Sustainable Farming Systems. Poster Papers* (C.E. Pankhurst, ed.), pp. 129-130. CSIRO, Melbourne, Australia.

Hodge, A., Stewart, J., Robinson, D., Griffiths, B.S. and Fitter, A.H.(2000) Plant N capture and microfaunal dynamics from decomposing grass and earthworm residues in soil. *Soil Biol. Biochem.*, 32: 1763-1772.

Holmstrup, M. (2000) Field assessment of toxic effects on reproduction in the earthworms *Aporrectodea longa* and *Aporrectodea rosea. Environ. Toxicol. Chem.*, 19: 1781-1787.

Holmstrup, M., Costonzo, J.P. and Lee, R.E. (1999) Cryoprotective and osmotic response to cold acclimation and freezing in freeze-tolerant and freeze- intolerant earthworms. *J. comp. Physiol.*, 169: 207-214.

Holmstrup, M., Jensen, K.S. and Kirknel, E. (1998) Estimation of effects of Benomyl application on the reproductive rate of earthworms in a grass field. In: *Advances in Earthworms Ecotoxicology. Proc. from the 2nd Int. Workshop on Earthworm Ecotoxicology* 2-5 April 1997. Amsterdam, The Netherlands. Pensacole FL. *Soc. Environ. Toxicol. Chem.*, pp. 353-364.

Hoogerkamp, M., Rogaar, H. and Eijsackers, H.T.P. (1983) Effects of earthworms on grassland on recently reclaimed polder soils in the Netherlands. In: *Earthworms Ecology- from Darwin to Vermiculture* (J.E. Satchell, ed.), pp. 85-105. Chapman and Hall, London, England.

Hopp, H. (1947) The ecology of earthworms in cropland. *Soil Sci. Soc. Am. Proc.*, 12: 503-507.

Huhta, V. and Koskenniemi, A. (1975) Numbers, biomass and community respiration of soil invertebrates in spruce forests at two latitudes in fiinland. *Ann. Zool. Fennici,* 12: 164-182.

Ismail, S.A. (1986) Earthworms resources of Madras. In: *Proc. Natl. Sem. Org. Waste Utiliz. Vermicomp.* Part B: *Verms and Vermicompositing* (M.C. Dash, B.K. Senapati and P.C. Mishra, eds.), pp. 8-15. Five Star Printing Press, Burla, India.

Ismail, S.A. (2000) Organic waste management.*Technology Appreciation Programme on Evaluation of Biotechnological Approches to Waste Management* held on 26th Oct, 2000. Industrial Association-Ship of IIT, Madras, pp. 28-30.

Ismail, S.A. and Murthy, V.A. (1985) Distribution of earthworms in Madras. *Proc. Indian Acad. Sci. (Anim. Sci.)*, 94: 557-566.

Ismail, S.A., Ramakrishnan, C. and Anzar, M.M. (1990) Density and diversity in relation to the distribution of earthworms in Madras. *Proc. Indian Acad. Sci (Anim. Sci.),* 99: 73-78.

Jager, T. (1998) Mechanistic approach for estimating bioconcentration of organic chemicals in earthworms (Oligochaeta). *Environ. Toxicol. Chem.,* 10: 80-90.

Jambhekar, H.A. (1990) Effect of vermicompost as a biofertilizer on grape wines VIIIth Southern Regional congerence on microbial inoculant, Pune, India.

James, S.W. (1991) Soil nitrogen, phosphorous and organic matter processing by earthworms in lallgrass prairie. *Ecology*, 72: 2101-2109.

Jensen, J. (1993) Applications of vermiculture technology for managing organic waste resources. *Proc. 9th Int. Conf. on Solid Waste Management*, 14-17November, 1993. pp 1-8.

Julka, J.M. (1988) *The Fauna of India and the Adjeacent Countries.* Megadrile: Oligochaeta (Earthworms). Haplotaxida: Lumbricina: Megascolecoidea: Octochaetidae, XIV+ 400 p. Zoological Survey of India, Calcutta.

Julka, J.M. (1993a) Earthworm resources of India and their utilization in vermiculutre. In: *Earthworm Resources and Vermiculture.* pp. 51-55 Zoological Survey of India, Solan.

Julka, J.M. (1993b) Distribution pattern in Indian earthworms. In: *Earthworm Resources and Vermiculture*, pp. 27-31. Zoological Survey of India, Solan.

Julka, J.M. (1995) Oligochaeta: *Himalayan Ecosystem Series: Fauna of Western Himalayan,* Part 1: *Uttar Pradesh. Zool. Surv. India,* pp. 17-22.

Julka, J.M. (1996) Annelid diversity in the Thar desert. In: *Faunal Diversity in the Thar Desert::Gaps in Research* (A.K.Ghosh, Q.H. Bagri and I. Prakash, eds.), pp. 71-76. Scientific Publishers, Jodhpur, India.

Julka, J.M. (2001) Distribution of earthworms in different agroclimatic regions of India. *Workship on Tropical Soil Biology and fertility programme.* School of Environmental Sciences, J.N.U., New Delhi, pp. 1-12.

Julka, J.M. and Mukherjee, R.N. (1984) Some observations on the seasonal activity of earthworms (Oligochaeta:Annelida) in hill forest soil. *Bull. Zool. Surv. India,* 5: 35-39.

Julka, J.M. and Mukherjee, R.N. (1986) Preliminary observations on the effect of *Amynthas diffringens* (Oligochaeta: Megascolecidae).

In: *Proc. Nat. Sem. Org. Waste Utiliz. Verimicomp.* Part B: *Verms and Vermicompositing* (M.C. Dash, B.K. Senapati and P.C. Mishra, eds.), pp. 66-68. Five Star Printing Press, Burla, India.

Julka, J.M. and Paliwal, R. (1986) Distribution of Indian earthworms. In: *Verms and Vermicomposting.Proc. Natl. Sem. Org. Waste Utiliz. Vermicomp.* Part B (M.C. Dash, B.K. Senapati and P.C. Mishra, eds.), pp.16-22. Five Star Printing Press, Burla, India.

Julka, J.M. and Paliwal, R. (1990) Seasonal changes in the population of earthworms (Oligochaeta) in an Orchard. *J. Bomb. Nat. Hist. Soc.,* 87: 323-326.

Julka, J.M. and Senapati, B.K. (1987) Records of the zoological survey of India. Miscellaneous Publication. Occ Pap. 92. Grafic Printall, Calcutta, India, pp. 1-105.

Kale, R.D. (1998) Earthworms: Nature's gift for utilization of organic wastes. In: *Earthworm Ecology* (C.A. Edwards, ed.), pp. 355-376. CRC Press, The Netherlands.

Kale, R.D. (2000) An evaluation of the vermitechnology process for the treatment of agro, sugar and food processing wastes. *Technology Appreciation Programme on Evalution of Biotechnological Approaches to Waste Management* held in 26th Oct. 2000. Industrial Association-Ship of IIT, Madras, pp. 15-17.

Kale, R.D. (2001) The functional role of earthworms in agroecosystems. Workshop on Tropical Soil Biology and Fertility Programme. School of Environmental Sciences, J.N.U., New Delhi, pp. 1-5.

Kale, R.D. and Bano, K. (1986) Field trials with vermicompost [Vee comp. E. 83 UAS] an oganic fertilizer. In: *Proc. Natl. Sem. Org. WasteUtilize. Vermicomp. Part B: Verms and Vermicomposting* (M.C. Dash, B.K. Senapati and P.C Mishra, eds.), pp. 151-156. Five Star Printing Press, Burla, India.

Kale, R.D. and Krishnamoorthy, R.V. (1978) Distribution of earthworms in relation to soil conditions in Bangalore. In: *Soil Biology and Ecology in India, UAS Tech. Ser.* (C.A. Edwards and G.K. Veeresh, eds.), pp. 63-69. University of Agricultural Sciences, Bangalore, India.

Kale, R.D. and Krishnamoorthy, R.V. (1981) What affects the abundance and diversity of earthworms in soil? *Proc. Indian Acad. Sci. (Anim Sci.),* 90: 117-121.

Kale, R.D. and Krishnamoorthy, R.V. (1982) Cyclic fluctuations in the population and distribution of the three species of tropical earthworms in a farmyard garden in Bangalore. *Rev. Ecol. Biol. Sol.,* 19: 61-71.

Kale, R.D. and Seenappa, S. N. (1997) Earthworm in agriculture. *Training Course on Organic Farming UAS*. Bangalore, pp. 1-5.

Kale, R.D., Mallesh, B.C., Bano, K. and Bagyaraj, D.J.(1992) Influence of vermicompost application on the available macronutrients and selected microbial populations in a paddy field. *Soil Biol. Biochem.*, 24: 1317-1320.

Kollmannsperger, F. (1956) Lumbricidae of humid and arid regions and their effect of soil fertility VI Congr. Inst. Sci. Sol. Rappp. C., pp. 293-307.

Kretzschmar, A. (1998) Earthworm interactions with soil organization. In: *Earthworm Ecology* (C.A. Edwards, ed.), pp.163-176. CRC Press, The Netherlands.

Krishnamoorthy, R.V. (1989) Factors affecting the surface cast production by some earthworms of Indian tropics. *Proc. Indian Acad. Sci. (Anim. Sci.)*, 98: 431-445.

Krishnamoorthy, R.V. and Vajranabhaiah, S.N. (1984) "An assessment of plant growth promoter levels in worm casts." *A paper presented to National Seminar on Organic Utilization and Vermicomposting*, School of Life Sciences, Sambalpur University, Orissa 5-8 Dec. 1984, 8 p.

Kubiena, W.L. (1953) Bestimmungsbuch and Systematik der Boden Europas, 392 p. Stuttgart.

Kubiena, W.L.(1955) Animal activity as a decisive factor in establishment of humus forms. In: *Soil Zoology* (D.K. Mc.E. Kevan, ed.), pp. 73-82. Butterworths Scientific Publications, London.

Kula, H. and Kokta, C. (1992) Side effects of selected pesticides on earthworms under laboratory and field conditions. *Soil Biol. Biochem.*, 24: 1711 –1741.

Lavelle, P. (1974) Les vers de terre de la savane de Lamto. In: *Analyse d'un Ecosysteme Tropicale Humide: la Savane de Lamto* (Cote d'Ivoire). *Bull. de Liaison des Chercheurs de Lamto* (Paris), 5: 133-136.

Lavelle, P. (1983) The structure of earthworm communities. In: *Earthworm Ecology-from Darwin to Vermiculture* (J.E. Satchell, ed.), pp. 449-466. Chapman and Hall, London, England.

Lavelle, P. (1988) Earthworms activities and the soil system. *Biol. Fertil. Soil,* 6: 237-251.

Lavelle, P. and Martin, A. (1992) Small-scale and large-scale effects of endogeic earthworms on soil organic matter dynamic in the humid tropics. *Soil Biol. Biochem.,* 24: 1491-1498.

Lavelle, P., Barois, I., Martin, A., Zaidi, Z. and Schaefer, R. (1989). Management of earthworm populations in agroecosystems: A possible way to maintain soil quality? In: *Ecology of Arable Land.* (M. Clarholm and L. Bergstrom, eds.), pp. 109-122. Kluwer, The Netherlands.

Lavelle, P., Dangerfield, M., Fragoso, C., Eschenbrenner, V., Lopez-Hernandez, D., Pashanasi, B. and Brussaard, L. (1994) The relationship between soil macrofauna and tropical soil fertility. In: *The Biological Management of Tropical Soil* (M. J. Swift and P. Woomer, eds.), pp. 137-169. John Wiley-Sayce, New York.

Lavelle, P., Pashanasi, B., Charpentier, F., Gilot, C., Rossi, J.P., Derouard, L., Andre, J., Francois, J.P. and Bernier, N.(1998) Large-scale effects of earthworms on soil organic matter and nutrient dynamic. In: *Earthworm Ecology* (C.A. Edwards, ed.), pp. 103-122. CRC Press, The Netherlands.

Lee, K.E. (1959) The earthworm fauna of New Zealand. *Bull. N. Z. Dep. Scient. ind. Res.*, 130: 1-486.

Lee, K.E. (1985) Earthworms: *Their Ecology and Relationships with Soils and Land Use*. Academic Press, London, 411 p.

Lee, K.E. (1992) Some trends and opportunities in earthworm research or Darwin's children- the future of our discipline. *Soil Biol. Biochem.*, 24: 1765- 1771.

Lee, K.E. and Foster, R.C.(1991) Soil fauna and soil structure. *Aust. J. Soil Res.*, 29: 745-775.

Linnaeus, C. (1758) Systema Naturae. *Regnum Animale,* 10: 824.

Ljungstorm, P.O. (1967) The origin of the Swedish earthworm fauna. In: *Progress in Soil Biology* (D. Graff, and J.E. Satchell, eds.), pp. 547-550. North Holland Publishing Co., Amsterden, The Netherlands..

Loehr, R.C., Martin, J.H., Neuhauser, E.F. and Malecki, M.R. (1984) Waste management using earthworms-engineering and scientific relationship. Final Report, Project ISP-8016764 (National Science Foundation, Washington), 118 p.

Lofes, H.A. (1980) Measuring growth of earthworms as a method of testing sublethal toxicity of pesticide: Experiment with benomyl and trichloroacetic acid. *Swedish J.Agric. Res.*, 10: 25-34.

Lofs- Holmin, A. (1986) Occurrence of eleven earthworm species (Lumbricidae) in permanent pastures in relation to soil-pH. *Swedish J. Agric. Res.*, 16: 161-165.

Madan, M. and Sharma, N. (1986) Recycling of organic wastes through vermicomposting. *Bioenergy Newsletter*, 2: 30-31.

Madge, D.S. (1969) Field and laboratory studies on the activities of two species of tropical earthworms. *Pedobiologia*, 9: 188-214.

Marinissen, J.C.Y. and Bok, J. (1988) Earthworm-amended soil structure: its influence on collembola populations in grassland. *Pedobiologia,* 32: 243-252.

Marinissen, J.C.Y. and de Ruiter, P.C. (1993) Contribution of earthworms to carbon and nitrogen cycling in agroecosystems. *Agric. Ecosys. Environ.,* 47: 59-74.

Marinissen, J.C.Y. and van den Bosch, F. (1992) Colonization of new habitats by earthworms. *Oecologia,* 91: 371-376.

Martin, A. (1991) Short- and long- term effects of the endogeic earthworm *Millsonia anomala* (Omodeo) (Megascolecidae, Oligochaeta) of tropical savannas, on soil organic matter. *Biol. Fertil. Soil,* 11: 234-238.

Maurya,N. and Chattoraj,A.N. (1994) Insecticidal interaction with a non-target soil organism-earthworm.In:*Soil Environment and Pesticides* (D.Prashad and H.S. Gaur, eds.), pp. 201-231. Venus Publishing House, New Delhi, India.

Mazaud, D. and Bouche, M.B. (1980) Introduction en surpopulation et migrations de lombriciens marques. In: *Soil Biology as Related to Land Use Practices* (D.L. Dindal, ed.), pp. 687-701. E.P.A. Washington.

Mclean, M.A. and Parkinson, D. (2000) Introduction of the epigeic earthworm *Dendrobaena octaedra* changes the oribatid community and microarthropod abundances in a pine forest. *Soil Biol. Biochem.*, 32: 1671-1681.

Mclnroy, D. (1971) Evaluation of the earthworms *Eisenia foetida* as food for man and domestic animals. *Feedstufls,* 43: 37-47.

Mekada, H., Hayashi, N., Yokota, H. and Olcomura, J. (1979) Performance of growing and laying chickens fed diets containing earthworms. *Jpn. poult. Sci.,* 16: 293-297.

Michaelsen, W. (1903) *Die Geographische Verbreitung der Oligochaetan.* R. Friedlander and Sohn, Berlin.

Michaelsen, W. (1933) Die Oligochaetenfauna Surinames mit Erarterung der verwandtschaftlichen Beziehungen der Octochatinen. *Tijdschr. Ned. Dierk. Vereen.*, 3: 112-131.

Mishra, K.C. and Ramakrishnan, P.S. (1988) Earthworm population dynamics in different Jhum fallows developed after slash and burn agriculture in north- eastern India. *Proc. Indian Acad. Sci. (Anim. Sci.),* 97: 309-318.

Mohamed, A.I., Nair, G.A., Kassem, H.H. and Nuruzzaman, M. (1995) Impacts of pesticides on the survival and body mass of the earthworm *Aporrectodea caliginosa* (Annelida: Oligochaeta). *Acta. Zool. Fennica,* 196: 344-347.

Muys, B. and Granval, P.H. (1997) Earthworms as bioindicators of forest site quality. *Soil Biol. Biochem.,* 29: 763-766.

Muys, B. and Lust, N. (1992) Inventory of the earthworms communities and the state of litter decomposition in the forests of Flanders, Belgium and its implications for forest management. *Soil Biol. Biochem.,* 24: 1677-1681.

Nakamura, Y. (1968) Studies on the ecology of terrestrial Oligochaeta, I. Seasonal variation in the population density of earthworms in alluvial soil grassland in Sapporo, Hakkaido. *Appl. Entomol. Zool.* 3:89-95.

Neuhauser, E.F., Loehr, R.C. and Malecki, M.R. (1988) The potential of earthworms for managing sewage sludge. In: *Earthworms and Waste Management* (C.A. Edwards and E.F. Neuhauser, eds.), pp. 9-20. SPB Academic Publishing, The Netherlands.

Neuhauser, E.F., Loehr, R.C., Malecki, M.R., Milligan, D.L. and Durkin, P.R. (1985) Toxicity of selected organic chemicals to the earthworm, *Eisenia fetida. J. Enviorn. Qual.*, 14: 383-388.

Neuhauser, E.F., Malecki, M.R. and Adnatra, M. (1986) Comparative toxicity of ten organic chemicals to four earthworm species. *Comp. Biochem. Physiol.,* 83: 197-200.

Norgrove, L. and Hauser, S. (2000) Production and nutrient content of earthworm casts in a tropical agrisilvicultural system. *Soil Biol. Biochem.,* 32: 1651-1660.

Norstorm, S. and Rundgren, S. (1973) Associations of lumbricids in Southern Sweden. *Pedobiologia,* 13: 301-326.

Oades, J.M. and Walters, L.J. (1994) Indicators for sustainable agriculture; Policies to paddock. In: *Soil Biota: Management in Sustainable Farming Systems* (C.E. Pankhurst, B.M. Doube, V.V.S.R. Gupta and P.R. Grace, eds.), pp. 219-223. CSIRO, Melbourne.

Olive, P.J.W. and Clark, R.B. (1978) Physiology of reproduction. In: *Physiology of Annelids* (J.Mill, ed.), pp. 683. Academic Press, London.

Olson, H.W. (1928) The earthworms of ohio, with a study of their distribution in relation to hydrogen- ion concentration, moisture and organic content of the soil. Ohio *Biol. Surv. Bull.*, 17: 47-90.

Pani, S.C. and Senapati, B.K. (1986) Effect of malathion on population dynamic, reproductive biology and secondary production of *Drawida willsi*, Michaelsen, from upland irrigated paddy field soil of Orissa, India. In: *Proc Nat. Sem. Org. Waste Utiliz. Vermicomp. Part B: Verms and Vermicomposting* (M.C. Dash, B.K Senapati and P.C. Mishra, eds.), pp. 97-100. Five Star Printing Press, Burla, India.

Parkinson, D. (1988) Linkages between resource availability microorganisms and soil invertebrates. *Agric. Ecosys. Environ.*, 24: 21-32.

Parkinson, D. and Mclean, M.A. (1998) Impacts of earthworms on the community structure of other biota in forest soils. In: *Earthworms Ecology* (C.A Edwards,ed.), pp. 213-226. CRC Press, The Netherlands.

Parmelee, R.W., Beare, M.H., Cheng, W., Hendrix, P. F., Crossley, D.A.Jr. and Coleman, D.C. (1990) Earthworms and enchytraeids in conventional and no-tillage agroecosystems: A biocide approach to assess their role in organic matter breakdown. *Biol. Fertil. Soil.*, 10: 1-10.

Perel, T.S. (1977) Differences in lumbricid organization connected with ecological properties. In: *Soil Organisms as Components of Ecosystems, Proc. 6th Int. Coll. Soil Zool. Ecol. Bull.* (Stockholm), 25: 56-63.

Petersen, S.O. and Holmstrup, M. (2000) Temperature effects on lipid composition of the earthworms *Lumbricus rubellus* and *Eisenia nordenskioeldi. Soil. Biol. Biochem.*, 32: 1787-1791.

Phillipson, J., Abel, R., Steel, J. and Woodell, S.R.J. (1976) Earthworms and the factors governing their distribution in an English beechwood. *Pedobiologia*, 16: 258-285.

Piearce, T.G. (1972) Acid intolerant and ubiquitous Lumbricidae in selected habitats in North Wales. *J. Anim. Ecol.*, 41: 397-410.

Prashad, B. (1916) The earthworms of Lahore. *J. Bombay Nat. Hist. Soc.*, 24: 494-506.

Reddy, M.V. (1988) The effect of caste of *Pheretima alexandri* on growth of *Vinca rosea* and *Oryza sativa*. In: *Earthworms in Environmental and Waste Management* (C.A., Edwards and E.F., Neuhauser, eds.), pp. 241-248. SPB Academic Publishing, The Netherlands.

Reddy, M.V. and Alfred, J.R.B. (1978) Some observations on the earthworm population and the biomass in a subtropical pine forest soil. In: *Soil Biology and Ecology in Indian, UAS. Tech. Ser.* (C.A. Edwards and G.K. Veeresh, eds.), pp. 78-82. University of Agricultural Sciences, Bangalore, India.

Reddy, R.D., Puruhotham, K.R. and Rammurthi, R. (1984) Effect of sublethal concentration of Sumithion (Fenitrothion) on carbohydrate metabolism in the South Indian earthworm *Lampito mauritii*. *J. Environ. Biol.*, 5: 119-123.

Reddy, V.R. and Reddy, M.V. (1990) Response of population structure and biomass of earthworms to conventional tillage in a semi-arid tropical grassland. *Soil Biol. Ecol.,* 10: 73-78.

Reinecke, A.J. and Reinecke, S.A. (1998) The use of earthworms in ecotoxicological evaluation and risk assessment: New approaches. In: *Earthworms Ecology* (C.A. Edwards, ed.), pp. 273-293. CRC Press, The Netherlands.

Reinecke, A.J. and Visser, F.A. (1980) The influence of agricultural land use practices on the population densities of *Allolobophora trapezoides* and *Eisenia rosea* (Oligochaeta) in southern Africa. In: *Soil Biology as Related to Land Use Practices* (D.L. Dindel, ed.), pp. 310-324. EPA, Washington.

Reinecke, A.J., Viljoen, S.A. and Saayman, R.J. (1992) The suitability of *Eudrilus eugeniae, Perionyx excavatus* and *Eisenia fetida* (Oligochaeta) for vermicomposting in southern Africa in terms of their temperature requirements. *Soil Biol. Biochem.,* 24: 1295-1307.

Reynolds, J.N. and Cook, D.W. (1993) Nomenclatura Oligochaetologica Supplementum Tertium, a catalogue of names, descriptions and type specimens of the Oligochaeta. *New Brunswick Mus. Monogr. Ser. (Nat. Hist.),* 9: 39.

Reynolds, J.W. (1972a) The relationship of earthworm (Oligochaeta: Acanthodrilidae and Lumbricidae) distribution and biomass in six heterogeneous woodlot sites in Tippecanoe country, Indiana. *J.Tennessee Acad. Sci.,* 47: 63-67.

Reynolds, J.W. (1972b) Earthworms (Lumbricidae) of the Haliburaton Highands, Ontairo, Canada. *Megadrilogica.*, 1: 1-11.

Reynolds, J.W. (1973) Earthworms (Annelida: Oligochaeta) ecology and systematics. In: *1st Soil Microcommunities Conf. Natl. Tech. Inform Serv.* (D.L. Dindal, ed.), pp. 95-120. Springfield, USA..

Reynolds, J.W. (1994) The distribution of the earthworms (Oligochaeta) of Indiana: A case for the post- Quaternary introduction theory for magadrile migration in North America. *Megadrilogica,* 5: 13- 32.

Reynolds, J.W. and Jordan, G.A. (1975) A preliminary conceptual model of megadrile activity and abundance in the Haliburton Highlands. *Megadrilogica*, 2: 1-9.

Reynoldson, T.B. (1955) Observation on the earthworms of North Wales. *North Wales Nat.*, 3: 291-304.

Robinson, C.H., Piearce, T.G., Ineso, P., Dickson, D.A. and Nys, C. (1992) Earthworm communities of limed coniferous soils: Field observations and implications for management. *Forest Ecol. Mgt.*, 55: 117-134.

Rovira, A.D., Smettem, K.R.J. and Lee, K.E. (1987) Effects of rotation and conservation tillage on earthworms in a red brown earth under wheat. *Aust. J. Exp. Agric.*, 31: 509-513.

Rundgren, S. (1977) Seasonality of emergencen lumbricids in southern Sweden. *Oikos*, 27: 476-482.

Rusek, J. (1985) Soil microstructures – contributions on specific soil organisms. *Quasetiones Entomologicae*, 21: 497-514.

Saetre, P. (1998) Decomposition microbial community, structure and earthworm effects along a birch-spruce soil gradient. *Ecology*, 79: 834-846.

Sahu, S.K., Mishra, S.K. and Senapati, B.K. (1988) Population biology and reproductive strategy of *Dichogaster bolaui* (Oligochaeta: Octochaetidae) in two tropical agroecosystems. *Proc. Indian Acad. Sci. (Anim. Sci.)*, 97: 239-250.

Sahu, S.K. and Senapati, B.K. (1986) Population density, dynamics, reproductive biology and secondary production of *Dichogaster bolaui* (Michaelsen) . In: *Proc. Natn. Sem. Org. Waste Utiliz. Vermicomp.* Part B (M.C. Dash *et al.*, eds.), pp. 29-46. Sambalpur Univ., Jyoti Vihar, Orissa, India.

Satchell, J.E. (1955) Some aspects of earthworm ecology. In: *Soil Zoology* (D.K.Mc.E. Kevan, ed.), pp. 180-201. Butterworths, London, England.

Satchell, J.E. (1967) Lumbricidae. In: *Soil Biology* (A.Burges and F. Raw, eds.), pp. 259-322. Academic Press, London, England.

Satchell, J.E. (1983) *Earthworms Ecology-from Darwin to Vermiculture.* Chapman and Hall, London, England, 495 p.

Savingy, J.C. (1826) Memoire sur les lombrics. In: *Analyse des travaux de 1' Academie royale des Sciences, pendent 1' annee 1821, partie physique.* (M. Cuvier and G. Baron, eds.) *Mem. Acad. Sci. Inst. Fr.(Hist.)*, 5: 176-184.

Scheu, S. and Parkinson, D. (1994a) Effects of earthworms on nutrient dynamic carbon turnover and microorganisms in soils from cool temperate forests of the Canadian Rocky Mountains-laboratory studies. *Appl. Soil Ecol.*, 1: 113-125.

Scheu, S. and Parkinson, D. (1994b) Effects of invasion of an aspen forest (Canada) by *Dendrobaena octaedra* (Lumbricidae) on plant growth. *Ecology*, 75: 2348-2361.

Schulz, E. and Graff, O. (1977) Zur Bewertung von Regenwarmmehl aus *Eisenia fetida* als Eiweissfuttermittel. Landbouw forschung volkenrode, 27: 216-218.

Scullion, J. and Malik, A. (2000) Earthworm activity affecting organic matter, aggregation and microbial, activity in soils restored after opencast mining for coal. *Soil Biol. Biochem.*, 32: 119-126.

Seastedt, T.R. (1984) The role of microarthropods in decomposition and mineralization processes. *Ann. Rev. Entomol.*, 29: 25-46.

Semenova, L.M. (1968) Morpho–functional peculiarities of integuments in earthworms. *Zool. Zh.*, 47: 1621-1627. (In Russian, English Summary)

Senapati, B.K. (1994) Science for villages and agrobased industry through vermiculture. Changing Villages, 13: 1-17.

Senapati, B.K. and Dash M.C. (1981) Effect of grazing on the elements of production in the vegetation and oligochaete components of a tropical pasture land. *Rev. Ecol. Biol. Sol.*, 18: 487-505.

Senapati, B.K. and Dash, M.C. (1984) Functional role of earthworms in the decomposer subsystem. *Trop. Ecol.*, 25: 54-73.

Senapati, B.K. and Dash,(1991) Energetic strategy and life table analysis of two peregrine Indian earthworms. In: *Advances in Management and Conservation of Soil Fauna* (G.K. Veeresh, D. Rajagopal and C.A. Viraktamath, eds.), pp. 677-681. Oxford and IBH Publishing Co. Pvt. Ltd., New Delhi

Senapati, B.K., Dash, H.K. and Dash, M.C. (1979) Seasonal dynamics and emergence pattern of a tropical earthworm, *Drawida calebi* (Oligochaeta). *Int. J. Invert. Rep.*, 1: 271-277.

Sims, W. (1980) A classification and distribution of earthworms, suborder Lumbricina (Haplotaxida: Oligochaeta). *Bull.Br. Mus. Nat. Hist. (Zool.)*, 39: 103-124.

Singh, J. (1997) Habitat preferences of selected Indian earthworm species and their efficiency in reduction of organic materials. *Soil Biol. Biochem* ., 29: 585-588.

Singh, J. and Rai, S.N. (1998) Potential of earthworms in sustainable agriculture. *Yojana* (November, 1998), pp 10-12.

Springett, J.A. (1983) Effects of five species of earthworm on some soil properties. *J. Appl. Ecol.*, 20: 865-872.

Springett, J.A. (1985) Effect of introducing *Allolobophora longa* (Ude) on root distribution and some soil properties in New Zealand pastures. In: *Ecological Interactions in Soils: Plants, Microbes and Animals* (A.H. Fitter, D. Atkinson, D.J. Read and M.B. Usher, eds.), pp. 330-405. Blackwell, Oxford.

Springett, J.A. (1992) Distribution of lumbricid earthworms in New Zealand. *Soil Biol. Biochem.,* 24:1377-1381.

Stein, A. Bekker, R.M., Blom, J.H.C. and Rogaar, H. (1992) Spatial variability of earthworm populations in a permanent polder grassland. *Boil. Fertil. Soil,* 14: 260- 266.

Steinberg, D.A., Pouyat, R.V., Parmelee, R.W. and Groffman, P.M. (1997) Earthworm abundance and nitrogen mineralization rates along an urban- rural land use gradient. *Soil Biol. Biochem.,* 29: 427-430.

Stenersen, H. (1979) Action of pesticides on earthworms. Part I. The toxicity of cholinesterase inhibiting insecticides to earthworms evaluated by laboratory tests. *Pesticide Science*, 10: 66-74.

Stenersen, J., Gilman, A. and Vardanis, A. (1973) Carbofuran: Its toxicity to and metabolism by earthworms *(Lumbricus terrestris). J. Agric. Food.Chem*., 21: 166-171.

Stephens, P.M., Davoren, C.W., Ryder, M.H. and Doube, B.M. (1993) Influence of the lumbricid earthworm *Aporrectodea trapezoides* on the colonization of wheat roots by *Pseudomonas corrugata* strain 2140R in soil. *Soil Biol. Biochem.,* 25: 1719-1724.

Stephens, P.M., Davoren, C. W., Ryder, M. H. and Doube, B. M. (1994a) Field experiments demonstrating the ability of earthworms to reduce the disease severity of *Rhizoctonia* on wheat. In: *Soil Biota: Management in Sustainable Farming Systems* (*Poster Papers*) (C.E. Pankhurst, ed.), pp. 19-20. CSIRO, East Melbourne, Australia.

Stephens, P.M., Davoren, C.W., Ryder, M.H. and Doube, B.M. (1994b) Greenhouse and field experiments demonstrating the ability of earthworms to reduce the disease severity of Take-all on wheat. In: *Soil Biota: Management in Sustainable Farming Systems (Poster Paper)* (C.E. Pankhurst, ed.), pp. 19-20. CSIRO, East Melbourne, Australia.

Stephens, P.M., Davoren, C.W., Ryder, M.H., Doube, B.M. and Correll, R.J. (1995) Field evidence for reduced severity of *Rhizoctonia* bave patch disease of wheat, due to the presence of the earthworms *Aporrectodea rosea* and *Aporrectodea trapezoides. Soil Biol. Biochem.,* 26:1495-1500.

Stephenson, J. (1930) The Oligochaeta, XVI + 978 pp. Clarendon Press, London.

Stockdill, S.M.J. (1982) Effects of introduced earthworms on the productivity of New Zealand pastures. *Pedobiologia*, 24: 29-35.

Stockdill, S.M.J. and Cossens, G.G. (1966) The role of earthworms in pasture production and moisture conservation. *Proc. New Zealand Grassland Association*, pp. 168-183.

Stolte, H.A. (1962) Oligochaeta. In: *Bronn's Klassen Und Ordnungen des Tierreichs,* 4: 891-1141. Geest and Portig, Leipzig.

Subler, S., Baranski, C.M. and Edwards, C.A. (1997) Earthworm additions increased short-term nitrogen availability and leaching in two grain- crop agroecosystems. *Soil Biol. Biochem.,* 29: 413-421.

Svendsen, J.A. (1957) The distribution of Lumbricidae in an area of Pennine Moorland (Moor House, Nature Reserve). *J. Anim Ecol.,* 26: 409.

Taboga, L. (1980) The nutritional value of earthworms for chicken. *Bri. Poult. Sci.,* 21: 405-410.

Temple- Smith, M.G., Kingston, T.J., Furlonge, T.L. and Garnsey, R.B. (1993) The effect of the introduction of the earthworms *Aporrectodea caliginosa* and *Aporrectodea longa* on pasture production in Tasmania. *Proc. 7th Aust. Agro.Conf. Adelaid*, pp. 373.

Tian, G., Olimah, J.A., Adeoye, G.O. and Kang, B.T. (2000) Regeneration of earthworm populations in a degraded soil by natural and planted fallows under humid tropical conditions. *J. Am. Soc. Sci. Soil,* 64: 222 –228.

Tiunov, A.V. and Scheu, S. (2000) Microbial biomass, biovolume and respiration in *Lumbricus terrestris* L. cast material of different age. *Soil. Biol. Biochem.,* 32: 265-275.

Tripathi, G., Rai, S.N. and Singh, J. (1995) Ecology and vermitechnology of some Indian earthworms. In: *Int. Conf. Sust. Agril. Environ.,* HAV Hisar, India, 92 p.

Vail, V.A. (1974) Contributions on North Amercian earthworms (Annelida) II. Observations on the hatching of *Eisenia foetida* and *Bimastos tumidus* (Oligochaeta: Lumbricidae). *Bull . Tall Timbers Res. Sta.,* 16: 1 –18.

Van Rhee, J.A. (1967) Development of earthworm populations in orchard soils. In: *Progress in Soil Biology* (O. Graff. and J.E.

Satchell, eds.), pp. 360-371. North- Holland Publishing Co., Amsterdarn, The Netherlands.

Velasquez, L., Ibanez, I., Herrera, C. and Oyarzun, M. (1991) A note on the nutritional evaluation of worm meal (*Eisenia foetida)* in diets for rainbow trout. *Animal products*, 53: 119-122.

Visser, S. (1985) Role of the soil invertebrates in determining the composition of soil microbial communities. In: *Ecological Interactions in Soil: Plants, Microbes and Animals* (A.H. Fitter, D.Atkinson, D.J. Read and M.B.Usher, eds.), pp. 297-317. Blackwell Scientific Publication, Oxford, Great Britain.

Wadia, D.N. (1973) Geology. In: *The Gazetteer of India* 1. Publication Division, Ministry of Information and Broadcasting. Government of India, New Delhi, pp. 117-162.

Wood, T.G. (1974) The distribution of earthworms (Megascolecidae) in relation to soils, vegetation and altitude on the slopes of Mt Kosciusko, Australia. *J. Anim. Ecol.,* 43: 87-106.

Yeates,G.W. (1981) Soil nematode populations depressed in the presence of earthworms. *Pedobiologia*, 22: 191-195.

Yoshida, M. and Hoshii, H. (1978) Nutritional value of earthworms for poultry feed. *Jpn. Poult. Sci.*, 15: 308-311.

Zajona, I. (1968) Rozbor fauny dazdoviek (Oligochaeta, Lumbricidae) lucnych biocenoz na okoli liptovskej Mary. *Acta Fac. Nat. Univ. Comenianae. Zool.*, 14: 19-32.

Zazonc, I. (1970) Earthworms synusiae of meadow stratocenoses and choriocenoses. *Acta. Zooltech. Univ. Agr. Nitra.,* 21: 203-211.

2

Dynamic Potentials of Vermiresources

G. Tripathi, P. Bhardwaj and V.B. Lal

Earthworms are known to inhabit the earth's soils since the Precambrian period about 600 million years. They are segmented invertebrates belonging to the phylum Annelida, class Clitellata and order Oligochaeta. Earthworms are generally classified into epigeic, endogeic and anecic depending upon different lifestyles (Bouche, 1977). Epigeic worms live in soil surface of 3-10 cm and feed on leaf litter. Endogeic worms live deep in the soil from 10-30 cm and feed on the humic materials and mineral matter, where as anecic worms can go very deep into soil upto 60-90 cm and form complicated burrows for their movements. The external abiotic parameters and the poor soil nutrients appear to be controlling factors for growth of earthworm population. The distribution of worms mainly depends upon the physiochemical characteristics of the soil such as temperature, pH, moisture, inorganic salts, organic matter, soil texture and aeration. The most suitable temperature which earthworms prefer is 10-35 ^{0}C, moisture 12-34%, pH of about 7 and C/N ratio 2-8 (Lavelle, 1974; Philipson *et al.*, 1976; Edwards and Lofty, 1977; Kale and Krishnamoorthy, 1981; Lee, 1985). Earthworms avoid drought and dry soils either by migrating to lower layer or by entering a state of diapause. Well aerated loose soil rich in organic matter support earthworms largely than the hard, clay and poorly aerated soils.

Earthworms are omnivorous animals but they are often selective in their food habits. They derive their nutrition from organic materials, bacteria, fungi, diatoms, algae, protozoa, nematodes and decomposing animals. Surface living earthworms feed on food materials selectively, while deep soil living worms ingest soil as such. The type and amount of food materials available influence the size of the earthworm population, species diversity, growth-rate and cocoon production. Nitrogen and manure rich diets help in rapid growth of cocoon production in earthworms. Earthworms are usually not self mating although they are hermaphrodite.

Earthworms are important creatures of soil biota, which effects on the soil structure, aeration and fertility. Several scientists have studied the taxonomy and potentials of earthworms. Earthworms have a remarkable variety and variability. The economic importance of earthworm cannot be ignore in the present days. They may be used as a soil improver, pollution bioindicator, pollution controller, detoxifer, compost manufacturer, soil stabilizer, land reclaimer and medicine producers. Functionally, earthworms may be classified into two groups: humus formers and humus feeders. Humus formers include surface dwelling worms which mainly feed on nearly 90% organic matter and 10% soil. They are generally red in colour. They are also known as detritivores worms. These worms are harnessed for vermicomposting. The other group includes deep burrowing worms which feed on 90% soil and 10% degraded organic matter. They are faint in colour and are useful in making the soil pares and distributing humus throughout the soil.

Earthworms are familiar to everyone and their importance are globally realized that earthworms can do wonderful job in the management of different pedoecosystems. To a common man worms are rather insignificant animals which generally come out on the soil surface during rain. The role of earthworms in increasing soil fertility was, however, known to even ancient farmers, but during the last 2 to 3 decades, the use of artificial fertilizers have faded their significance. In recent years the farmers are once again realizing the benefits of these soil creature and are making all efforts to culture them (Jairajpuri, 1993). Many people call them as environmental monitoring tool because they accumulate heavy metals, industrial effluents, agricultural chemicals and their residues. Earthworms are very useful in land reclamation, soil improvement and organic waste management. In other word, earthworms may be referred as waste controller, compost manufacturers and protein producers (Singh and Rai, 1998).

The diversified potentials of earthworms can be discussed under the following headings:

(a) Waste decomposer

(b) Biofertilizer manufacturer

(c) Land reclaimer

(d) Protein producer

(e) Food source

(f) Drug source

(g) Vitamin source

(h) Natural detoxicant

(i) As a bait

Waste Decomposer

Pollution due to increase in the waste materials, especially with the growth of human population and industrialization, is one of the major hazards for the maintenance of clean environment. These environmental problems associated with agriculture have emphasized investigation on earthworms. The role of earthworms in transformation of waste into agriculturally and environmentally beneficial producer is not new rather earthworms have served our earth and mankind in this capacity for millions of year.

The waste materials can be processed under vermicomposting programme which involve the harnessing of earthworms for the stabilization of a variety of organic wastes. This is the most important aspect of earthworm biotechnology at present. Earthworms such as *Eudrilus eugeniae, Eisenia fetida* and *Perionyx excavatus* are easily adaptable to agricultural wastes like sugarcane thrash, paper pulp, faceal matter of cow, sheep, horse, activated sludge and biogas sludge of poultry dropping. The broken product of waste materials or the degraded organic matter, by worms activity is known as Vermicompost. These days sewage sludge and animal slurries are useful as soil conditioners and fertilizers due to their organic constituents and high levels of inorganic nutrients. The presence of earthworms in a sludge produces increased oxygen consumption, decreased anaerobic decomposition and increased mineralisation (Mitchell *et al.,* 1980). Many of these effects stimulate microbial activity by tunnelling and casting of the earthworms, which increase aeration and surface area of the sludge (Hartenstein and Hartenstein, 1981).

Unfortunately, little is known about the suitability of other earthworms species for vermicomposting. The few species which have been tested are found to be inferior to *Eisenia fetida* in terms of growth and development rates (Kaplan *et al.,* 1980). More recent work on ecology have shown the precise importance of earthworms in the cycling of organic matter in terrestrial ecosystem. The life cycle and ecology of *Eisenia fetida* (tiger worm) have therefore been the subject of intensive research. Two cocoons are produced at each mating upto 20 ova are present in each cocoon. Incubation time varies with temperature but may be in short as 2-3 weeks under controlled conditions (Stephenson, 1930). Wide temperature fluctuations are tolerated by this species.

The contribution of earthworms to soil fertility through their burrowing and feeding activities is a well recognized fact (White, 1789). Processing of organic waste in large quantities has been carried out in Toronto (Canada), Portland, La Jolla, San Diego, simi valley, Santa Raa valley and Los Angeles in the United States, (Riggle and Holmes, 1994). Several researchers and project developers have started discussing the role of earthworms in organic recycling. Earthworms are also used in New Zealand, United Kingdom, Spain and several other countries in organic waste recycling and in soil management. The advantages of using earthworms are many. They not only solve the problem of garbage disposal but at the same time provide excellent compost for application to gardens and fields, a benefit when the entire world is looking towards sustainable agriculture and organic farming.

Biofertilizer Munufacturer

Low level of soil fertility is one of the major problems enhancing the agricultural production in India. The low fertility of soil has increased the fertilizers which in turn hit the small and marginal farmers. The major manurial resources in India are crop residues, tree and agricultural wastes, industrial byproducts, marine wastes and biofertilizers. The organic wastes create, environment problem like occupying vast area, spreading foul odours and forming breeding home for most of the pathogenic microorganisms and mosquito vectors. Earthworms are probably one of the major contributors in the breakdown of organic materials. They also have the capacity to accumulate several inorganic substances including heavy metals. Earthworms ingest fresh or partially decomposed organic matter from the soil surface. The ingested organic matter is broken into small pieces by the grinding action of the earthworm gizzard. The fragmented organic matter is mixed with enzymes in the earthworm digestive tract, and excreted as a colloidal humus which is rich in plant nutrients. Under such condition, the only alternative for providing a renewable supplementary source nutrients for crop plants in the country is the recycling of wastes. Organic recycling is important for the maintenance of soil sustainability. The use of compost to replace the chemical fertilizers has become necessary (Madan and Sharma, 1986).

Vermicomposting helps to process wastes simultaneously giving biofertilizers and proteins. Earthworms eat and mix large amount of soil and organic matter then deposit their casts either on the soil surface or in burrows depending on species. Their cast contain high concentration of organic material, slit, clay and cations such as iron, calcium, magnesium and potassium. Earthworms also release nitrogen into soil in their casts and in the mucus. The exchangeable cations such as Ca, Mg, Na, K, available P and Mo in wormcasts is more than in the surrounding soil (Shinde *et al.,* 1992). Dash (1978) suggested that the worms along with the organic manures can be utilized as an alternative to costly inorganic fertilizers for growing crops. Tomati *et al.* (1983) reported that earthworm cast which is rich in available nutrient increased the plant growth and yield of crop.

Animals, vegetables and industrial wastes can be processed to produce the organic fertilizers at sites without much extra cost (Edwards, 1988). In order to get a good environment, more cultivation and sustainable agriculture, earthworms are introduced in agricultural fields for providing nutrient rich vermifertilizers. Species of earthworms such as *Eudrilus eugeniae, Eisenia fetida* and *Perionyx excavatus* are easily adaptable to various organic waste materials and produces vermicompost which can be used as the organic manure in the field to prevent organic carbon deficiency and soil erosion. Earthworm's dead tissues are added to the soil as nitrogenous fertilizers. Earthworm casting are also used as casing layer for mushroom cultivation.

Kale and Bano (1986) made field trails of vermicompost and found it as effective as chemical fertilizers. The production and application of vermifertilizers have many advantages. This can be well examplified from the vermiculture technology programme of Cuba (Rosset and Benjamin, 1993). In Cuba vermicomposting programme has started in 1986, with two small boxes of red worms. This country established 172 vermicomposting centers upto 1992 which produce 93,000 tons of worm humus. It has been demonstrated that four tons of vermicompost can replace the use of forty tons of cow manure per hectare by giving 36% higher yield. Vermicomposting appear to be the most promising which helps to process wastes, simultaneously giving biofertilizer for agricultural and horticultural uses.

The report on specific feeding interaction between *Eisenia fetida* and microorganism has led to the suggestion that certain types of industrial wastes, which are regarded as unsuitable for disposal by earthworms, may be seeded with appropriate microbes to render them adaptable to vermicomposting. A wide range of animal wastes are also suitable for vermicomposting. The development in this field suggests that anaerobically digested animal wastes could be utilized as a substrate for earthworm growth. This results in digested residues which has fertilizer value.

Biological conditioning of wastes through vermitechnology has many advantages (Fig.1). The products of vermicomposting are earthworm biomass and wormcast. The major product of vermicomposting is the earthworm casting which acts as plant growth promoter. The significant improvement in plant growth which have been obtained using vermicomposts suggests that further work on quality control and product variability may be worthwhile, in view of the greater financial returns to be expected from sales to the horticultural market.

The recent development of vermicomposting at Pune institute found burrowing earthworms, *Pheretima elongata*, especially efficient in breaking down the toughest of organic wastes like sugarcane thrash. The work at Banaras Hindu University Varansi supported by United States India Fund (WSPP) project has identified *Amynthas morrisi, Dichogaster bolaui* and *Perionyx sansibaricus* as potential decomposers. Now the solid waste planners and environmental engineers are looking with great interest at the potentials of worms to convert wastes into assets.

Land Reclaimer

In many parts of the world, agricultural productivity is lowered by decreased soil fertility or poor physical conditions. Uncultivated land may be reclaimed for agriculture use. Earthworms play a significant role in management and reclamation of degraded pedoecosystem and enrich the barren soil with nutrients. The introduction of earthworm populations in the exhausted soils make it more compact and its poor structure changes into deep friable top soil, this in turn is beneficial for plant growth and crop yield. In many areas suitable earthworms species has a significant positive effect upon nutrient cycling. In New Zealand, it has been shown that earthworms can reduce surface run-off from pasture slopes and water erosion by increasing the water soaking capacity of soils.

Charles Darwin observed the activities of earthworms for about 40 years and wrote in his famous book (1881) "The Formation of Vegetables Mould through Action of Worms" about earthworms as a major contributor in increasing soil fertility. The role of earthworms in decomposition process has been well documented (Edwards and Lofty 1977; Satchell, 1983; Senapati and Dash, 1984; Lee, 1985; Christensen, 1988; Reinecke *et al*, 1992; Vincesias Akpa and Loquet, 1997; Singh, 1997). Julka and Mukherjee (1986) demonstrated the potentiality of earthworm in decomposition of organic matter of sterilized soil. Spiers *et al.* (1986) reported the importance of indigenous earthworms in decomposition and nutrient cycling in the forest ecosystems of Canada. Work at the State University of New York in USA has shown that earthworms can breakdown activated sludge very effectively (Harstenstein *et al.*, 1979). The Worms Concern in California has a capacity to cycle more than 100 tons waste per day.

The introduction of *Aporrectodea caliginosa* doubles the rate of water infilteration and increases the amount of water stored in the soil by 70%. *Aporrectodea longa* is known for improving soil fertility and production when introduced to agricultural fields. For this reason scientists are now considering spreading of this all over the Australia. *Lumbricus terrestris* is specialist at creating channels deep into the soil that increase the capacity of plant to grow deep roots. The plants can then take up more water and nutrients than in surface soil establishing worms as a part of sustainable form management. Earthworms has the potential to increase fertility and production of plants on many Australian forms.

Stockdill (1959) have successfully introduced earthworms into pastures in New Zealand that lacked native earthworms. Some of the methods of earthworms inoculation for the improvement and reclamation of land have been tried by various workers. These methods includes direct release of worm, as soil blocks, as turf pieces and as vermicasting. Barley and Kleining (1964) successfullty introduced *Aporrectodea caliginosa* and *Microscolex dubuis* into newly sown, irrigated pasture on sandy loam in Australia with significant improvement in soil structure, loses of organic materials and increased productivity. Van Rhee (1969, 1971) introduced earthworms into polders that had been drained and reclaimed from the sea in the Netherlands. He found an increase in development of normal soil structure and return of the polders to productive use.

The success of the land or soil reclamation depends on identification of proper species suitable for different agroclimatic zones of a country and their threshold numbers. The researches on this aspect are urgently required to improve the quality of exhausted or degraded soils in our country. In Indian condition *Polypheretima elongata* is suitable for soil reclamation. This species produces more mucus which binds available nutrients. It helps in nutrient restoration and moisture conservation. (Singh and Rai, 1998).

The rehabilitation of open-cast mining sites and industrial spoil heaps may be done by introducing earthworms. The presence of earthworms rapidly improves water balance, organic matter availability and microbial activity, and hence facilitates the establishment and growth of vegetation. The use of earthworms for soil and land reclamation is likely to increase in developed, developing and undeveloped countries as the problem of population and land use become more acute.

Protein Producer

Earthworms are also used as protein rich sources of animal feed. Earthworm contain 70-80 percent protein on a dry weight basis and this protein is of a high quality. Analysis have shown that it has a good balance of essential amino acids and is especially rich in lysine. The amino acid composition of earthworms is far superior to snail and fish meat. The essential amino acids spectrum of earthworm tissue is very rich than the currently used sources of feed proteins. The presence of essential amino acids in earthworms tissue are sufficient in order to fulfill the recommendations of FAO/WHO particularly in terms of lysine, methionine, cysteine and tyrosine all of which are important components of animal feed (Singh and Rai, 1998). Lawrence and Miller (1945) suggested that earthworms contained sufficient protein and considered as animal food. The species of *Perionyx, Eudrilus, Metaphire* and *Eisenia* are good for biomass production. *Lampito mauritii* has high protein contents in their body tissue and they are suitable as fish bait, poultry and fish feed. Not only protein but high content of nitrogen and fat have also been reported in this species.

Perionyx excavatus can easily be cultured and can also provide animal protein for utilization in poultry and fish feed (Julka and Senapati, 1987). Guerro (1981) found *Perionyx excavatus* as a better animal protein source for *Tilapia* than local fish meal. Arunachalam and Palanichamy (1984) have shown the improved growth rate of fish *Mystus vittatus* when fed on dry or fresh worms. A small white earthworm (*Enchytraeus albidus*) is often grown in soil and used to feed aquarium fish and small laboratory animals.

Food Source

The tissues of earthworms contains various biochemical constituents. In addition to protein, the other useful biochemical constituents are fat 6-10%, carbohydrates 5-21%, minerals 2-3% and vitamins. After suitable processing, earthworms may be used as a food for livestock and aquaculture industry. They can even replace the traditional food. There are also reports that earthworms at some places are used for human consumption (Julka, 1988). They are a very promising source of high quality animal protein. A small white earthworm *Enchytraeous albidus* is often grown in soil and used to feed aquarium fish and small laboratory animals. They are eagerly eaten by some birds mainly robins and chickens. A large number of earthworms are eaten by frogs, moles, lizards, small snakes, centipeds and other predatory vertebrates and invertebrates. A lot of research is in progress in the country to serve worm meal as a protein rich supplement to poultry, aquaculture and domestic animals.

Drug Source

Earthworms are known to be associated with medicines since ancient time to cure various human diseases. In India, the paste of dried worms was prepared for curing diseases in Unani system of medicine. It has been used in treating wounds, Chronic boils, piles, sore throat, hernia and impotency when applied externally. When used internally they are useful in curing, chronic cough, diphtheria. jaundice, rheumatic pains, tuberculosis, bronchitis, facial paralysis and impotency. They have been used in folk medicine to treat pyorrhoea and small pox and to enable mothers to nurse their children in Burma, and for hair growth and expulsion of stones from the bladder in Iran (Stephenson, 1930). Hori *et al.* (1974) have reported the presence of antipyretic substances in tissue extracts of *Lumbricus spencer* and *Perichaeta communissuina*. The extracts may cure fever when treated in rabbit, in which fever have been induced by injection of *E. coli pyrogen*. It has been suggested that the antipyretic activity due to all cis-5, 8, 11, 14-ico satetraenoic acid (archiclonic acid) and all cis-5, 8, 11, 14, 17 – icosatetraaenoic acid. Presence of such a long chain fatty acids in earthworms do not affect the economics of vermiculture. Physiologically important compounds have also found in higher organisms like various monoamines (Gardner and Cashin, 1975).

There is a agreeing historical evidence of the value of earthworms in curing rheumatism (Reynolds and Reynolds, 1972). Even in these days the Chinese, the Japanese and the Indians are said to use earthworms in various fancy medicines. In Japan four kinds of drugs are known to be prepared from earthworms including antibioses, aphrodisics, antipyretics and antidots. Earthworms are being utilized in medicinal research to study the regeneration mechanism, cancer and birth control at some universities like California of USA (Mishra, 1984).

Vitamin Source

Earthworms are not only the source of proteins but they also have an excellent range of vitamins. The important vitamins present in the earthworm tissues are niacin, riboflavin (B2), pantothenic acid (B-complex), thiamine (B1), pyridoxine (B6), vitamin B12, folic acid and biotin (B-complex). Among these niacin and B12 are of significant value. Niacin is a valuable component of animal feeds (Edwards, 1985).

Natural Detoxicant

Due to rapid growth of population and industrialization, the rural and urban wastes are increasing continuously. These wastes are undesirable pollutants for the environment. Soil is being polluted by indiscriminate spray of pesticides. Release of toxic heavy metals and lethal radioactive materials are adversely affecting life in the soil. Earthworms form a major component of soil fauna due to their sensitiveness to such toxic and lethal materials. They can serve as convenient bioindicator of polluted soils (Julka, 1988). The pollutants are mainly solid wastes containing heavy metals released from the industries and pesticides used for health and agriculture. In recent years with the development of more and more industries, the amount of metals like mercury, lead, arsenic, cadmium etc. are released in the environment. These metals progressively reach the human being through the food chains. The special features of heavy metal chemicals are strong attraction to biological tissue but their slow elimination from biological systems. The accumulation of toxic chemicals in earthworm tissue is ecologically important, because they are the important component in the food chain of several species of birds and mammals. They form food for wood cock, robins and reptiles (Mishra, 1984) Earthworms have the ability to take up into their tissue a number of unwanted organic and inorganic chemicals. So pollutants can be detected from earthworms tissues which provide a good tool to test the level of pollution including industrial one. (Tripathi and Singh, 2000).

Many workers have described earthworms as bioindicator of soil contamination with heavy metals and pesticides, other agricultural chemicals acid rain etc. (Bouche, 1981; Paoletti *et al.,* 1991) Earthworms have been described as bioindicator of forest site quality (Muys and Granval, 1997). Ireland (1975) observed that earthworm *Drawida rubida* living in heavy metals contaminated acid soil can increase the amount of available Zn and Cu by excretion in the faeces with little or no effect on the amount of available lead.

Earthworms are selected as key indicator organisms for ecotoxicological testing of industrial chemicals. Conventional chemicals methods of analysis of pollutants are costly and give no indication of the bioavailability of material present in the environment. Alternative methods for this purpose have been proposed by the European Economics Community (EEC), OECD and also by FAO. Results indicate that earthworms are useful in assessing the presence of heavy metals pesticides, radioactive wastes and other toxic chemicals in soil and in screening of novel agricultural chemicals for their toxicity in the soil fauna.

As a Bait

It has been well demonstrated that the utilization of annelidian animal as a bait for fish market or for fish capturing is not only economically cheaper than the other traditional fish baits but also easily available. Fish capturing is not only a option of food source but also a very interesting entertainment. As we know in our rural areas and far from costal region of sea, the utilization of earthworms is used as very good fish bait on large scale. The earthworm bait market have been extending rapidly due to shortage of traditional fish bait such as shrimps and coarse fish during the past few years.

Earthworms are the best bait for fish as compared to other baits due to following important reasons:

1. Easy availability in every climate and pedoecosystems of tropical, subtropical and temperate regions.
2. Easy to rear and culture at wide scale without extra burden of money, equipment's, space and at low tech level.
3. Earthworm biomass and vermicompost are the extra output of application of vermitechnology for management of vegetable waste materials which create many ecological problems. Thus earthworm biomass can be used for fish bait and simultaneously we can manage organic wastes.

Now the earthworm fish bait market have been established at large scale in many developed countries. Exchange of millions dollar have been done by this earthworm based fish markets. Therefore, it can play valuable role in sustainable development of economic of a country. During 1980, about 500 million Canadian night crawlers (*Lumbricus terrestries*) were exported to USA. Its costs were estimated as 175 million at the rate of US$35 per thousand worms.

Summary

The growth of industries and ever increasing human population are changing the structure and function of soil communities. This also results in accumulation of waste materials. It in turn reduces the fertility of soils. All these problems which are associated with agriculture have prompted studies on earthworms. Due to their burrowing and casting activities the earthworms turn over much of the soil from bottom to the top. Darwin calculated that earthworms bring up about 18 tons of soil per acre per year. The excretory wastes of earthworms enrich the exhausted soil. The burrows of earthworm allows air and water to easily penetrate the soil. Inoculation of earthworms in orchards has been claimed to increase fruit production. They are used as bait for fishing. Earthworms are referred to as environmental bioindicator, waste controller, composed manufacturer and protein producer. Popularization of vermiculture biotechnology (VBT) will not only increase the production of low cost earthworm proteins but also play a significant role in pollution control.

REFERENCES

Arunachalam, S. and Palanichamy, S. (1984) Earthworms as feed for the catfish *Mystus vittatus*. In: Proc. *Natl. Sem. Org. Waste Utiliz. Vermicomp.* 4 p. School of Life Sciences, Sambalpur University, Jyoti Vihar, Orrisa.

Barley, K.P. and Kleining, C.R. (1964) The occupation of newly irrigated land by earthworms. *Aus. J.Sci.*, 26: 290.

Bouche, M.B. (1977) Strategies lombriciens. *Ecol. Bull.,* 25: 122-132.

Bouche, M.B. (1981) Development et lombriciences: Biostimulation des soils et bio-indicator. In: *Compte rendu des Journees Science ecologigue et development.*, pp. 281-295. AFIE, Grenoble.

Christensen, O. (1988) The direct effects of earthworms on nitrogen turnover in cultivated soils. *Ecol. Bull.*, 39: 41-44.

Darwin, C. (1881) *The Formation of Vegetables Mould through the Action of Worms with Observations on their Habits.* 326 p. John Murray, London.

Dash, M. C. (1978) Role of earthworm in the decomposer system. In: *Glimpses of Ecology* (J.S. Singha and B. Gopal, eds.), pp. 399-406. International Scientific Publication, India.

Edwards, C.A. (1985) Production of feed protein from animal wastes by earthworms. *Phil. Trans. Res. Soc. London*, 310: 299-307.

Edwards, C.A. (1988) Breakdown of animal, vegetable and industrial organic wastes by earthworm. In: *Earthworms in Waste and Environmental Management* (C.A. Edwards and E.F. Neuhauser, eds.), pp. 21-31. SPB, The Hague.

Edwards, C.A. and Lofty, J.R. (1977) *Biology of Earthworms.* 333 p. Chapman and Hall, New York.

Gardner, C. R. and Cashin, C. H. (1975) Some aspects of monoamine function in the earthworm (*Lumbricus terresrtris*). *Neuropharmacol.*, 14: 493-500.

Guerro, R. D. (1981) The culture and use of *Perionyx excavatus* as protein resource in the Philippines. In: *Proc. Darwin Centenary Symp.* 22 p.. Grange-over-sands, Great Britain.

Hartenstein, R. and Hartenstein, F. (1981) Physiochemical changes effected in activated sludge by the earthworm *Eisenia fetida*. *J. Environ. Qual.*, 10: 377-382.

Hartenstein, R., Neuhauser, E. F. and Kaplan, D.L. (1979) A progress report on the potential use of earthworms in sludge management. In: *Proc. 8th Natl. Sludge Conf. Silver Springs.* pp. 238 – 241. Maryland, USA.

Hori, M., Kondon, K., Yoshida, T., Konishi, E. and Minami, S. (1974) Studies of antipyretic components in the Japanese earthworm. *Biochem. Pharmacol.*, 23: 1583-1590.

Ireland, M .P. (1975) The effect of the earthworm *Dendrobaena rubida* on the solubility of Zn, Pb and Ca in heavy metal contaminated soils in water. *J.Soil Sci.*, 26: 313 – 318.

Jairajpuri, M. S. (1993) Earthworms and Vermiculture – An Introduction. *Earthworm Resources and Vermiculture.* pp. 1-5. Zoological Survey of India, Calcutta, India.

Julka, J.M. (1988) *The Fauna of India and the Adjeacent Countries.* Megadrile: Oligochaeta (Earthworms). Haplotaxida: Lumbricina: Megascolecoidea: Octochaetidae, XIV+ 400 p. Zoological Survey of India, Calcutta, India.

Julka, J.M. and Mukherjee, R.N. (1986) Preliminary observations on the effect of *Amynthas diffringens* (Oligochaeta: Megascolecidae). In: *Proc. Natl. Sem. Org. Waste Utiliz. Verimicomp.* Part B: *Verms and Vermicompositing* (M.C. Dash, B.K. Senapati and P.C. Mishra, eds.), pp. 66-68. Five Star Printing Press, Burla, India.

Julka, J.M. and Senapati, B.K. (1987) Records of the Zoological Survey of India. Miscellaneous Publication. Occ Pap. 92. pp. 1-45. Grafic Printall, Calcutta, India.

Kale, R.D. and Bano, K. (1986) Field trials with vermicompost [Vee comp. E. 83 UAS] an oganic fertilizer. In: *Proc. Natl. Sem. Org. WasteUtilize. Vermicomp. Part B: Verms and Vermicomposting* (M.C. Dash, B.K. Senapati and P.C Mishra, eds.), pp. 151-156. Five Star Printing Press, Burla, India.

Kale, R.D. and Krishnamoorthy, R.V. (1981) What affects the abundance and diversity of earthworms in soil? *Proc. Indian Acad. Sci. (Anim Sci.)*, 90: 117-121.

Kaplan, D.L., Hartenstein, R., Neuhauser, E.F. and Malecki, M.R. (1980) Physicochemical requirements in the environment of the environment *Eisenia foetida. Soil Biol. Biochem.*, 12: 347-352.

Lavelle, P. (1974) Les vers de terre de la savane de Lamto. In: *Analyse d'un Ecosysteme Tropicale Humide: la Savane de Lamto* (Cote d'Ivoire). *Bull. de Liaison des Chercheurs de Lamto* (Paris), 5: 133-136.

Lawrence, R.D.and Miller, R.H. (1945) Protein content of earthworms. *Nature*, 3939:517.

Lee, K.E. (1985) Earthworms: *Their Ecology and Relationships with Soils and Land Use*. 411 p. Academic Press, London.

Madan, M. and Sharma, N. (1986) Recycling of organic wastes through vermicomposting. *Bioenergy Newsletter*, 2: 30-31.

Mishra, P.C. (1984) Soil pollution and earthworm. In: *Proc. Natl. Sem. Org. Waste Utiliz. Vermicomp.*, pp. 24-29. School of Life Sciences, Sambalpur University, Orissa, India.

Mitchell, M. J. and Horner, S. G. (1980) Decomposition processes in sewage sludge and sludge amended soils. In: *Proc. 7th Intl. Colloq. Soil Zool.*, New York, USA. pp.129-138. Office of Pesticide and Toxic substance, EPA Washington, D. C.

Muys, B. and Granval, P.H. (1997) Earthworms as bioindicators of forest site quality. *Soil Biol. Biochem.*, 29: 763-766.

Paoletti, M. G., Favretto, M. R., Stinner, B.R., Purrington F. F. and Bater, J. E. (1991) Invertebrates as bioindicators of soil use. *Agric. Ecosyst. Environ.*, 34: 341 – 362.

Phillipson, J., Abel, R., Steel, J. and Woodell, S.R.J. (1976) Earthworms and the factors governing their distribution in an English beechwood. *Pedobiol.*, 16: 258-285.

Reinecke, A.J., Viljoen, S.A. and Saayman, R.J. (1992) The suitability of *Eudrilus eugeniae, Perionyx excavatus* and *Eisenia fetida* (Oligochaeta) for vermicomposting in southern Africa in terms of their temperature requirements. *Soil Biol. Biochem.,* 24: 1295-1307.

Reynolds, J. N. and Reynolds, W. M. (1972) Earthworms in medicine. *Amer. J. Nursing.* 72: 1273.

Riggle, D. and Holmes, H. (1994). The role of earthworms in organic recycling. *Biocycle.* 35: 58-62.

Rosset, P. and Benjamin, M. (1993) Soil Conservation: A key to the new model. In: *Two Steps Back, One Step Forward: Cuba's Nationalwide Experiment with Organic Agriculture,* pp. 38-50. Global Exchange, California.

Satchell, J.E. (1983) *Earthworms Ecology-from Darwin to Vermiculture,* 495 p. Chapman and Hall, London, England.

Senapati, B.K. and Dash, M.C. (1984) Functional role of earthworms in the decomposer subsystem. *Trop. Ecol.,* 25: 54-73.

Shinde, P.H., Naik, R. L., Nazirkar, R.B., Kandam, S.K. and Khaire, V.M. (1992) Evaluation of vermicompost. In: *Proc. Natl. Sem. Org. Farming*, pp. 54 – 55. MPKV, Pune, India.

Singh, J. (1997) Habitat preferences of selected Indian earthworm species and their efficiency in reduction of organic materials. *Soil Biol. Biochem* ., 29: 585-588.

Singh, J. and Rai, S.N. (1998) Potential of earthworms in sustainable agriculture. *Yojana* (November, 1998), pp 10-12.

Spiers, G. A., Gagnon, D., Nason, G.E., Packee, E.C. and Lousier, J.D. (1986) Effect and importance of indigenous coastal forest ecosystems. *Can. J. For. Res.*, 16: 983 – 989.

Stephenson, J. (1930) The Oligochaeta, XVI + 978 pp. Clarendon Press, London.

Stockdill, S.M.J. (1959) Earthworms improve pasture growth. *New Zealand J. Agric.,* 98: 227-233.

Tomati, V., Grapelli, A.,Galli, E. and Rossi, W. (1983) Refertilizer from vermiculture as an option for organic waste recovery. *Agrochimica*, 27:244-251.

Tripathi, G. and Singh, J. (2000) Earthworm biodiversity and vermitechnology in industrial waste management. In: *Industry Environment and Pollution* (P.K. Goel and A.Kumar, eds.), pp. 317-343, ABD Publishers, Jaipur, India.

Van Rhee, J.A. (1969) Inoculation of earthworms in a newly drained polder. *Pedobiol.*, 9: 128-132.

Van Rhee, J.A. (1971) Some aspects of the productivity of orchards in relation to earthworm activities. *Annal. Zool. Ecol.*, (special publication) 4: 99-108.

Vinceslas, A.M. and Loquet, M. (1997) Organic matter transformations in lignocellulosic waste products composted or vermicomposted (*Eisenia fetida andrei*): Chemical analysis and ^{13}C CPMAS NMR spectroscopy. *Soil Biol. Biohem.*, 29: 751-758.

White, G. (1789) The natural history of antiquities of selbourne in the country of Southampton. In: *Resources and Applications of Biotechnology* (R. Greenshields, ed.), pp. 49. Stockton Press, New York.

3

Vermiculture Biotechnology

S.C. Talashilkar, R.G. Kadam,
A.A. Todkari, P.V. Inamdar
and R.V. Dhopavkar

Earthworm, a member of the under ground zoo beneath our feet is Nature's own tiller, aerator, crusher, composter, moisture conservator, master builder of the top soil and above all is soil's intimate friend and benefactor. Biodegradation of rural residues, household garbage, sewage sludge and wastes from agro-industries by earthworms is an attractive proposition to regenerate valuable organic fertiliser consistent with minimisation of environmental pollution. Vermiculture is a source of good quality protein and medicine. It is, therefore, high time to encourage vermitechnology in the country with integrated approach by earthworm producers, processors and consumers.

Introduction

Earthworms, the soil invertebrates, alongwith soil micro-organisms carry out a yeomen's service of degrading organic waste materials and thus maintain the nutrient flux in the system. Even though the earthworms craftsmanship for formation, development and conservation of soil is known from the time of Darwin during late nineteenth century, vermitechnology as such has not been realised before midtwentieth century. The various countries that are engaged in the technology are U.S.A., Mexico, West Germany, Italy, Holland, Switzerland, Austria, Japan, Philippines and Thailand.

Biological conditioning of wastes through vermi-technology has multiple advantages as depicted in Fig. 1.

Fig. 1: Vermi technology flow sheet

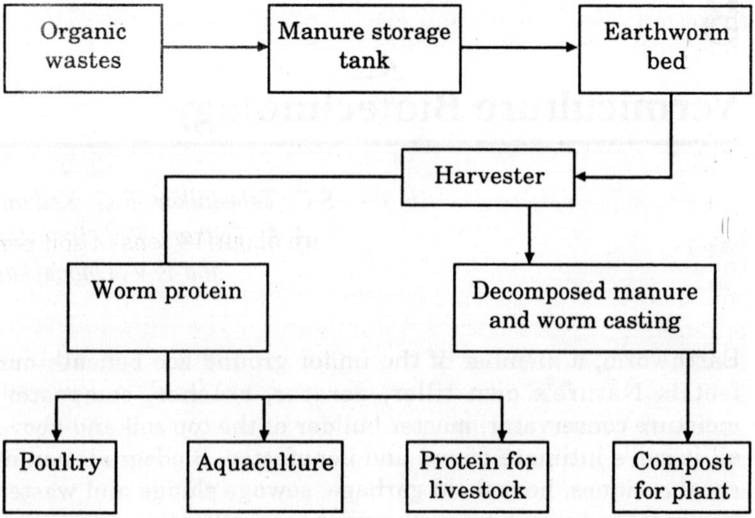

Earthworms as a Potential Source for Biodegradation

Charles Darwin (1881) first studied the role of earthworms in breaking down of dead plant and animal residues in soil. Later on many of the research workers studied the mechanisms of conversion of organic matter into the humus by introducing vermiculture in the field. They have shown the importance of the earthworms for decomposition of organic wastes such as pig and cattle solids and sludge, wastes from layers, chickens, broilers, turkeys and ducks, horse manure and rabbit droppings, crop residues and city garbage (Mackay *et al.*, 1962; Edwards, 1983; Sharma and Madan, 1983; Tomiti *et al.*, 1983; Senapati *et al.*, 1984; Talashilkar, 1986; Edward *et al.*, 1985; Bano and Kale, 1992 and Jambhekar, 1992).

Types of Earthworms

Earthworms are known to be inhabiting the earth's soils since the precambrian dating about 600 million years, bearing a silent witness to plant and animal evolution through the several landmarks and debacles of the evolutionary paradox. The report on the finding of fossilised earthworm cocoons (Piearce *et al.*, 1990) lends creditability to the ancestry. The basic design of the earthworms has not changed much over these millions of years and also does not vary much between the species.

Earthworms are simple, cylindrical, coelomate and segmented animals belonging to the order Oligochaeta of the Phylum Annelida. They are characterized by the presence of setae in their body segments, the setae being modified in their genital segments called differently as the pencil setae or copulatory setae. Earthworms form a major component of the soil biota and they together with a large number of other organisms constitute the soil community. The chief source of food to the soil biota is the litter contributed by plants. Although the dead plant tissues constitute the bulk of the food ingested by the earthworms, living microorganisms, fungi, microfauna and mesofauna and their dead tissues are also ingested as an important part of the diet (Parle, 1963a, b; Piearce, 1978). Though earthworms are generally called as saprophages, they can be classified based on the feeding habits (Lee, 1985) into *detritivores* and *geophages*.

Detritivores feed at or near the soil surface, mainly on plant litter or dead roots and other plant debris in the organic matter-rich surface soil hoirzon or on mammalian dung. These worms are generally called as *humus feeders* and comprise of the *endogeic* worms. *Octochaetona thurstoni* is one such earthworm commonly available in Madras.

A different classification has been proposed by Bouche (1977) laying stress on ecological strategies. He classified earthworms into *epigics, anecics and endogeics*. The epigeics have no effect on the soil structure as they generally cannot dig. They are efficient agents of commination and fragmentation of leaf litter. These are broadly classified as *phytophagous* earthworms. The anecics feed on the leaf litter mixed with the soil of the upper horizons. They may also produce surface casts. These are called as *geophytophagous* earthworms.

Varieties of Earthworms

Earthworms are scientifically classified as animals belonging to the order Oligocheaeta, class Chaetopoda, phylum Annelida. In this phylum there are about 1800 species of earthworms grouped into five families and distributed all over the world. The most common worms in North America, Europe and Asia belong to the family Lumbricidae which has about 220 species. Earthworms range from a few millimeters long to over 1 meter, but most of the common species are 5-10 cm in length. Only a few types are of commercial importance.

Nightcrawlers: This earthworm is common to the northern states and may be picked from fields and lawns at night for commercial fish-bait sale. Although very popular with fishermen, they are not commonly raised on a commercial basis because they reproduce slowly and require special production and control procedures.

Field worms (also known as garden worms): These make excellent fish bait and are often preferred by those who want a small number of worms for their own use. They are not prolific breeders, so are not recommended for commercial enterprises.

Manures worms (also known as bandlings, red wigglers or angleworms because of their squirming reactions when handled). They are particularly adaptable to commercial production and are one of the two types most commonly grown by successful worm farmers.

Red worms are basically another types of manure worm, differing mainly in size and colour from their larger and darker cousins. They are also very adaptable to commercial production and together with manure worms constitute about 80 to 90 per cent of commercially produced worms.

Manure and red worms can adapt to living in many different environments. They will eat almost any organic matter at some stage of decomposing organic residues in the soil or residues and plant litter on the soil surface. Strong muscles mix the swallowed material and pass it through the digestive tract as digestive fluids containing enzymes and secreted and mixed with the materials. The digestive fluids release amino acids, sugars and other smaller organic molecules from the organic residues (which include living protozoa, nematodes, bacteria, fungi and other microorganisms as well as partially decomposed plant and animal materials). The simpler molecules are absorbed through intestinal membranes and are utilized for energy and soil synthesis. Earthworms lack specialized breathing devices. Respiratory exchange occurs through the body surface.

Reproduction: Earthworms are usually not self-mating although they are hermaphrodite (each individual possess both male and female reproductive organs). A mutual exchange of sperm occurs between two worms during mating. Mature sperm and egg class and nutritive fluid are deposited in cocoons produced by the clitellum, a conspicuous, girdle like structure near the anterior end of the body. The ova (eggs) are fertilized by the sperm cells within the cocoon, which then slips off the worm and is deposited in or on the soil. The eggs hatch after about 3 weeks, each cocoon producing from two to twenty baby worms with an average of four.

Distribution and Ecology of Earthworms

Generally the earthworms are classfied into epigeic, endogeic and anecic. The worms live in the soil surface of 3 to 10 cm and feed on organic matter like leafletter or animal excrements are called epigeic. They are very active and have high regenerative capacity within a short period of time. Normally they are richly pigmented worms. The endogeic worms live deep in the soil from 10 to 30 cm. Most of the endogeic worms feed on the humic materials and mineral matter. They have very long life cycles with limited regeneration capacity and lightly pigmented. The Anecic worms can go very deep into soil up to 60 to 90 cm and form complicated burrows for their movements. They plaster the burrows with their own excrements and the mucus secretions. The external abiotic parameters and the poor nutritive resources of the soil system appear to be the controlling factors for earthworm population.

The distribution of worms in soil is influenced by several factors of which the soil texture and aeration, temperature, moisture, pH, inorganic salts, organic matter, dung, litter, reproductive potential and dispersive power of the species are important. The temperate soils have higher percentage of earthworms, which feed directly on organic matter, and low percentage of humas feeders. In tropical countries, the humas feeders predominate over the organic matter feeding worms. The *Lumbricids* and *Pheretima posthuma* distributed all over the world are called peregrine.

The abundance and soil distribution of worms depend on the general soil characteristics. Well aerated loose soils rich with organic matter support worms largely than the hard, clay and poorly aerated soils. Worms move to collect food and to escape unsuitable environmental conditions and predators, to look for a sexual partner. They also move into the soil or come to the surface to deposit their casts. Movement will be most rapid when the medium is smooth or porous. During this process, earthworms mix the soil, aerate and improve its water holding capacity.

Earthworms are very sensitive to hydrogen ion concentration. Many species of earthworms prefer soils with a pH of about 7. However *L. terrestris* occurs in soil with a pH of 5.4 whereas *Megascolex* thrives well in soils with pH ranging from 4.5 to 4.7. *Eisenia foetida* dominates in soils of pH 7 to 8. *B. eiseni*, *D. exatedra* and *D. rutrida* are acid tolerant species and *A. caliginosa*, *A. nocturna*, *A. chlorotica*, *A. longa* and *A. rosea* are acid intolerant species. Soil pH may also influence the worms that go into diapause.

Soil moisture is also an important factor in the distribution of earthworms. Earthworms avoid drought and dry soils either by migrating to lower layers or by entering a state of diapause. Majority of worms occurred in soils where the moisture content ranged from 12 to 45 per cent. *A. caliginosa* survived well at moisture content varying from 15 to 34 per cent. The water content of *E. foetida* and *P. huperensis* varied from 82 to 85 per cent of body weight. A moisture content of 23 per cent favoured *H. africans* to produce casts. Generally earthworms are more

Table 1: Niche diversification in earthworms

Earthworm species	*Weight adult worms*	*Temperature tolerance*	*Moisture tolerance*	*Feeding niche and virtical distribution*	*Active phase*	*Casting activity*	*Distribution*
1	2	3	4	5	6	7	8
Eudrilus eugeniae	1.5-2.5	18-35	20-40	Eplgeic, surface living in organic matter	Throughout the year, No diapause	Non-burrower, surface casting,	Tropical Africa and South America
Eisenia foetida	0.3-0.7	15-30	20-40	Epigeic, Living organic matter	Throughout the year, No diapause	Non-burrower, surface casting, loose, granular	Temperate regions of Europe, North America, India
Perionyx excavatus	0.8-1.2	8-30	30-50	Epigeic, Living in organic matter	Throughout the year, No diapause	Non-burrower, surface casting, loose, granular	Tropical countries
Dichogaster caurgensis	0.07-0.12	20-25	20-30	Partially epigeic surface living in litter and soil with high humus content	July-October, No resting stage	Non-burrower, surface casting, loose, granular	South India at high altitudes and heavy rainfall area

(Table Contd...)

Table 1: (Contd...)

1	2	3	4	5	6	7	8
Dichogaster bolaui	0.04-0.07	20-28	20-30	Partially epigeic, surface living in litter. Survive even in mineral soil with poor organic matter	July-October, No resting stage	Non-burrower, surface casting, loose, granular	Tropical countries
Dichogaster affinis	0.04-0.07	20-28	20-30	Partially epigeic, surface living in mineral soil with poor organic matter	July-October, No resting stage	Non-burrower, surface casting, oose, granular	Tropical countries
Drawida barwelli	0.2-0.5	20-30	40-50	Endogeic, 3-10 cm living humus and mineral soil	August-November seasonal cessation	Surface casting, caste are thick long, mostly underground	Tropical countries shade essential for establishment
Lampito maurit	0.8-1.5	18-28	20-40	Endogeic, 6-40cm living in humus and mineral soil	June-August seasonal cessation	Surface casting, caste loose and granuar complicated burrow	Plains of Indian Penninsula
Octochaetona	1.2-2.0	20-28	30-50	Endogenic, 3-10 cm living	August-November		

Govindan (1998)

active in moist soil than dry soil. The abundance of *A. caliginosa* and *L. rubellus* is correlated with soil temperature, moisture and supply of plant residue. *Lampito mauritii* and *Pontoscolex corethrurus* are abundant in soil where the C/N ratio was high but *Perionyx excavatus* is found commonly in soil with low C/N ratio. Earthworm species and their environmental conditions are given in Table 1. The species of earthworms identified in the soil of Konkan are given in Table 2.

Table 2: Species of earthworms identified from Konkan region of Maharashtra

Sr. No.	*Name of the species*	*Family*
1.	*Hoplochaetella khandalaensis* (Stephenson)	Octochaetidae
2.	*Hoplochaetella kempi* (Stephenson)	Octochaetidae
3.	*Perionyx sp.*	Megascolecidae
4.	*Megascolex konkanensis* (Fedrab)	Megascolecidae
5.	*Metaphire houlleti* (Perrier)	Megascolecidae
6.	*Amynthas alexandri* (Beddard)	Megascolecidae
7.	*Pontoscolex corethrurus* (Muller)	Glossoscolecidae
8.	*Drawida kanarensis* (Stephenson)	Megascolecidae

Food Habits of Earthworms

Earthworms are omnivovous animals but often selective in their food habits. Earthworms derive their nutrition from organic materials, living bacteria, fungi, diatoms, algae, protozoa, nematodes and decomposing animals. Surface living earthworms feed on food materials selectively while deep soil living worms ingest soil as such. The kind and amount of food materials available influence the size of the earthworms population, species diversity, growth rate and cocoon production. Activated sludge, manure and nitrogen rich diets help in rapid growth and more cocoon production in earthworms. The consumption rates of earthworms are given in Table 3.

In general earthworms are highly resistant to many pesticides and heavy metals. Worms could overcome the effect by increasing mucus secretion, restricting the movements and increasing the reproductive potential upto certain concentration levels. Earthworms are also known to concentrate the pesticides and heavy metals in their tissues. Sodium was concentrated by *Eisenia foetida* but not lead. In the United States studies have shown that. Earthworms concentrate cadmium, cobalt, mercury and zinc. Mature adult of *L. rubellus* had higher levels of cadmium than had unmature adults. The bioaccumulation of toxicants in the tissues of earthworms varies depending on the soil properties, pH, calcium concentration and so on.

Culturing of Earthworms

Culturing of earthworms is to be done in doors in humid places with proper shelter to avoid direct sunlight or heavy downpour. The culturing can be done in vermary, cattle shed or any other congenial place. Cement tanks, wooden boxes or plastic trays of approximately 1 m x 1 m x 0.3 m size, which accommodate 1000 worms may be used as containers. These should be free from cracks and holes to avoid escape of worms. Suitable covers should be provided to protect the worms from predators and prevent them from escaping during nights. For the bedding materials, sawdust or husk, fine sand or layer of garden soil are used. Three centimeters of each of the materials is to be spread in the container. Water is to be sprinkled on the bed to get a moisture level of 40-50 per cent. After the preparation of bed, cocoons or worms can be introduced into the containers.

The dung of cattle, sheep, horses, pigs or droppings of poultry birds and vegetable waste form the ideal feed for the worms. Cattle dung can be fed as such whereas the other

Table 3: Consumption rate of earthworms

Earthworms species	*Consumption rate mg substrate/g fresh wt. of worms/day*	*Substrate*	*Reference*
Allolobophora caliginosa	80	Dung	Barley, 1961
Allolobophora caliginosa	200-300	Soil	Barley, 1961
Allolobophora longa	20	Soil	Satchell, 1967
Eisenia foetida	10-5000	Activated sludge	Mitchell, 1978
Eisenia foetida	200-20000	Activated sludge	Hartenstein et al., 1979
Eudirirlus eugeniae	2000-5000	Activated sludge with dead leaves	Jeyashankar, 1993
Eudirirlus eugeniae	3000-7000	Activated sludge mixed with garbage	Balaji, 1994
Lampito mauritii	700-2800*	Soil	Dash et al., 1980
Lumbricus rubellus	27	Litter	Franz and Leitenberger, 1948
Lumbricus terrestris	27-80	Elm leaves	Needham, 1957
Lumbricus terrestris	10-30	Soil	Satchell, 1967
Millsonia anomala (Adult)	4000-7000	Soil	Lavele et al., 1980
Millsonia anomala (Young)	6000-18000	Soil	Lavele et al., 1980
Octolasium sp.	29	Soil	Crossley et al., 1971

* mg of substrate/g dry wt. of worms/day

dung material or vegetable wastes are to be mixed in equal quantity with cattle dung for the feed acceptability. Wheat bran, gram bran and vegetable waste when added to dung in 10:1:1:1 ratio would enhance the biomass production considerably. The results of the experiment conducted on conversion of poultry wastes into worm protein established that the poultry droppings and cow dung in 1:1 proportion are an appropriate diet in mass cultivation of earthworms for their protein production (Bano and Kale, 1984).

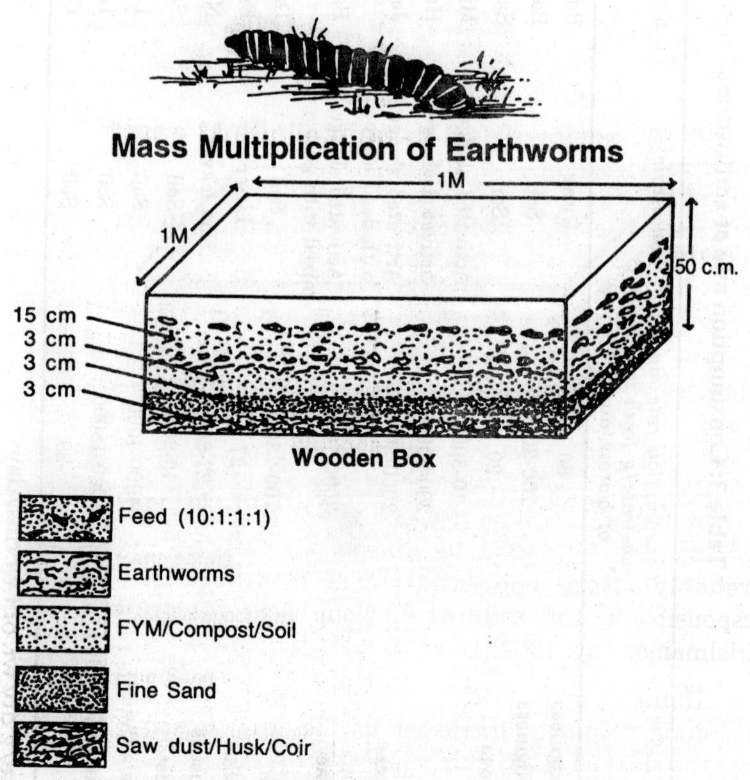

The dried food after pounding is to be mixed with the additives and water to form a loose dough. Then, the feed is to be placed towards a side on the culture bed. Worms beings surface feeders, aggregate beneath the feed and devour the feed. The feed is to be provided in small quantities at a time and replenished as and when it disappears from the surface. The worm cast in the containers should be brushed aside as and when left by the worms before providing fresh food.

Food Preference by Earthworms

The mechanism of eating of broad leaf litter by some earthworms was studied by Lindwist (1941). A survey of abundance and diversity of three earthworm species in and around Bangalore in relation to carbon and nitrogen content and humic and fulivic acid content conducted by Kale and Krishnamoorthy (1981) revealed that abundance of worm species differ with C/N and H/F ratios of the soil organic matter. *Lampito mauritii* and *Pontoscolex correthrurus* are abundant in soils with high C/N ratio, whereas *Perionyx exacavatus* is found in soils rich in nitrogen (low C/N ratios). Similarly, *L. mauritii* and *P. corethrurus* occupy lowly humified soils in contrast to the other species compared, optimal levels of C/N and H/F content can also determine the diversity of species in soils. The earthworm, *Lampito mauritii* showed preferences to leaf litter. Under a multiple choice test the worms preferred to settle in a medium containing mango, rice or ragi to any other leaf powder mixed with agar in cold water. Under no choice, the artificial agar feed containing mango leaf powder was consumed more by the worms. The preferences vary with reference to dry leaf matter and to disintegrating leaf in soil. Carbon and nitrogen of the leaf is not the criterion for this apetitive and consummatory behaviour. It is discussed that probably volatile potentiators or modifiers in leaf are responsible for the feeding behaviour of the worms (Kale and Krishnamoorthy, 1981).

Hatanbe *et al.* (1983), tested sludged cake, rice straw, cow dung manure, newspaper for the growth and observed that the sludged cake could support the growth of earthworms without processing to compost and when supplemented with cellulose material such as rice straw proved to be good growth medium. They also observed that the cow manure was also the best medium for growth of worms. Edward *et al.* (1985) mentioned that *Eisenia foetida* break down organic wastes which are rich in available nutrients and with good moisture holding capacity and porosity. These have considerable potential in horticulture as a plant growth medium. Kale and Bano (1988) carried out studies on dietary influence on the population of the *Eudrilus eugeniae* and reported that mix dung and grass gave higher population of 1476 number than other combination of diet. Patekar (1992) studied the multiplication rate of local species of earthworms (*P. corethrurus*) found in Konkan region in different wastes in combination with cow dung (Table 4).

Manna *et al.* (1994) studied the decomposition of waste materials such as wheat straw, maize stalk, chickpea straw, soybean straw and city garbage using three species of epigeic earthworms such as *L. mauritii*, kinberg; *Metaphire houlleti* and *Perionyx* spp. They observed appreciable increase in population of earthworms with the advancement of composting process upto 180 days in all the materials except in soybean straw.

Types of Earthworms Suitable for Vermicomposting

Some earthworms utilise microorganisms in their substrate as a food source and are able to selectively digest certain microorganisms (Dash, *et al.*, 1979; Cooke and Luxton, 1980 and Day, 1980). Biology of earthworm species suitable for composting is described by Gunathilagraj and Alfred (1996) (Table 5). The efficiency of vermicomposting system may, therefore, depends upon the number and types of microorganisms present in the substrate. Several workers have investigated the utilization of natural decomposition processes including earthworms in organic waste disposal. The species *E. foetida* is considered suitable for such application, because of its rapid growth rate, reproductive potential and occurrence in rich organic substrates in nature (Mitchell *et al.*, 1977; Neuhauser *et al.*, 1980 and Pincince *et al.*, 1980). Results of four months experiment showed that the manure samples inoculated with *E. foetida* decomposed more rapidly and showed a higher degree of humification than uninoculated samples (Riffaldi and Leviminzi, 1983). Chan and Griffiths (1988) studied the vermicomposting of pretreated pig manure in Hongkong using the earthworm species, *E. foetida*. The results showed

Table 4: Multiplication rate of local species of earthworms (*P. corethrurus*) in farm and kitchen waste

Sr. No.	*Diets*	*Initial worms introduced*	*Average No. of worms after*			
			20 days	*40 days*	*60 days*	*80 days*
1.	75% farm waste + 25% cowdung	10	14	28	58	89
2.	50% farm waste + 50% cowdung	10	15	28	62	99
3.	25% farm waste + 75% cowdung	10	12	12	38	91
4.	100% farm waste	10	11	13	29	58
5.	75% kitchen waste + 25% cowdung	10	14	24	59	89
6.	50% kitchen waste + 50% cowdung	10	11	21	57	90
7.	25% kitchen waste + 75% cowdung	10	12	17	65	94
8.	100% kitchen waste	10	10	13	25	56
9.	Control (soil)	10	12	16	29	42

that fecund earthworm species such as *E. foetida* is suitable for biorecycling pretreated pig manure, the worms grew rapidly and reproduced a humus rich worm cast which was odour free. From her experimental studies and commercial production, Jambhekar (1992) concluded that crop residues and animal wastes can safely be recycled through vermicomposting. The noticed hastening of the decomposition process and compost was ready within one and half months period.

India has about 3000 species of earthworms which are adopted to a range of vermiculture needs. Earthworms can be divided into the following two broad categories:

(i) Epigeic – the surface living worms

(ii) Epianecic – the burrowing worms

Table 5: Biology of earthworm species suitable for composting

	Eisenia foetid	*Eudrilus eugeniae*	*Perionyx excavatus*
Duration of life cycle (days)	± 70	± 60	± 46
Growth rate (mg worm^{-1} day^{-1})	7	12	3.5
Maximum body mass (mg individual worm)	1500	4294	600
Maturation attained at age (days)	± 50	± 40	± 21
Start of cocoon production (days)	± 55	± 46	± 24
Cocoon production (worm^{-1} day^{-1})	0.35	1.3	1.1
Incubation period (days)	± 23	± 16.6	± 18.7
Mean number of hatching (cocoon^{-1})	2.7	2.7	1.1
Number of hatching from one cocoon	1-9	1-5	1-3

Source: Gunthilagraj, K. and S. Alfred 1996.

Epigeic or manure worms are found on the surface and are reddish brown, they do not process the soils but are efficient in composting of organic wastes. They enhance the area of organic manure production through biodegradtion or minerlization and nutrient mobilization (Senapati, 1984). When maintained in captivity under semi-natural conditions, they remain active throughout the year. At Madras, local species being used are *L. mauritii, Octachoetona serrata* and *P. excavatus* (Ismail, 1994). Unlike the temperate regions, humus feeders (epianecic) dominate over the organic matter feeding (epigeic) worms in the tropical countries. In the tropics, a small group of four to ten cm long worms having dark pigmentation are found in the litter heap, dung pad or near the cattle shed during monsoon. The pigmentation, quick response to stimuli and high regeneration capacity enable their survival in unstable environment.

Methods of Vermicomposting

In the treatment of organic wastes, surface feeding earthworms function as a shredding machine, breaking up large lumps of the material as they ingest it, mixing the material and increasing the surface area to allow better aeration and drying. This in turn, stimulates the activities of decomposing microorganisms. The earthworms gut is a stable environment. The end result should be a fine, stable, odour free material, containing most of the nutrients from the original waste. In Asia, vermicomposting is being efficiently practiced in the Philippines. The setps followed as narrated by Gaur (1983) are as follows: The pits of 4 m x 3 m x 1 m size are dug and floor of pit is covered with a lattice of wood strips to provide drainage. The pit is filled with organic residues including straw, stubbles, animal manure, green weeds, leaves and so on. The filled pit is covered loosely with soil and kept moist. The worms are placed on the top, which burrow down in the damp soil. The pits are shaded from hot sunshine and to keep the material moist within two months about 10 kgs of casting are produced per kg of worms. Bhawalkar (1991) has suggested a new method of bio-processing of the residue generated from sugarcane trash, pressmud, sugar waste water, ash, spent wash etc. His package of methods consist of applying a basal dose of vermicastings @ 5 tonnes/ha and covering with a 25 mm layer of feed and a 100 mm layer of mulch of sugarcane trash, weed etc. Haines and Oren (1991) suggested breaking down of human and animal wastes by using *E. foetida*. In England, in the 1980's, following research of Rothamsted experimental station over some years, a company called British Earthworm Technology was formed to commercialise the work and developed a market for worm compost derived from animal wastes. In France, a project is successfully using earthworm to breakdown municipal organic waste from a large town. Similar projects are commenced in Germany which combine composting to decompose the waste to a certain stage and then introduce earthworm to produce a better quality compost. Bhawalkar Earthworm Research Institute, Pune has developed through twelve years of sustained field scale research, a cost effective large scale application of vermiculture biotechnology. This technology is being harnessed to set up units for cost effective treatment of various nontoxic organic solid and liquid waste from the cities, dairies, sugar and distillary units, pulp and paper mills, tanneries, fermentation industries and food processing units. Apart from environmental protection through the pollution abatement, these units serve a production centres of vermicastings which are being used as earthworm culture because of the beneficial microflora and worm cocoon associated with them (Bhawalkar and Bhawalkar, 1992).

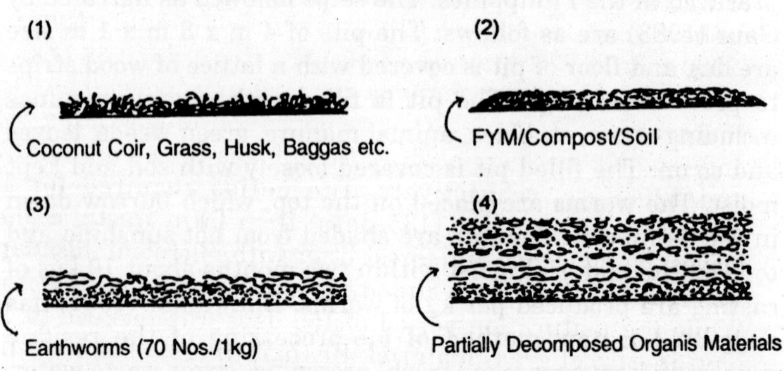

The following tips have been suggested to increase worm population and therefore, to boost up vermicompost production (Bhidey, 1994).

(a) A mixture of cattle, sheep, horse dung with gram, wheat bran and vegetable waste form ideal feed for worms.

(*b*) Mixing of gram bran with dung mixture in 3:2 ratio increases biomass. Mixing of wheat bran to dung mixture in 3:10 ratio hastens the growth of worms. Addition of kitchen waste in the same proportion also increases the worm population.

(*c*) The biogas sludge and poultry droppings in equal quantities enhance the worm population and biomass. A method of vermicomposting conducted at Dnyan Prabhodini, Pune involves preparation of soil layer of 15-20 cm thick at the bottom of the pits of 2 m x 1 m x 1 m followed by vermibed of fresh cattle dung. It is than layered with dry leaves or hay to about 5 cm. The moisture content of the pit without flooding is maintained through the addition of water. After four weeks, the organic wastes are spread over to a thickness of 5 cm at an interval of 3-4 days. At maturation, the moisture content be brought down for getting fine loose granular mass (Gaur and Geet Singh, 1995).

Jadhav (1996) conducted a field experiment on composting of wastes such as cow dung, rice straw, grass, mango leaves, cashew leaves and forest litter with and without earthworms (2000 numbers/heap) in the heaps of 2 m x 1 m x 1 m (Fig. 2). The results on the changes in various physico-chemical characteristics revealed that humification of cow dung with and without earthworms of *E. foetida* was completed after 90 and 120 days, respectively. The period required for the maturation of compost (135 days) from farm waste, cashew leaves, mango leave and forest litter with microbial inoculants was reduced by 30 days due to the inoculation of earthworms. While the maturation period of 150 days was reduced to 120 days in case of grass material. Bhangrath (1996) also studied the changes in properties like C/N ratio, water soluble carbohydrates, cation exchange capacity, humus fractionation etc. during humification of said materials carried out by pit method for a period of 150 days (Table 6). An inoculation of earthworms such as *E. foetida* and *E. eugeniae* resulted into reduction in the period of compost maturati on the kitchen garbage from 90 to 60 days. The depth of compost pit varying from 30 cm and 60 cm did not influence the earthworm activity.

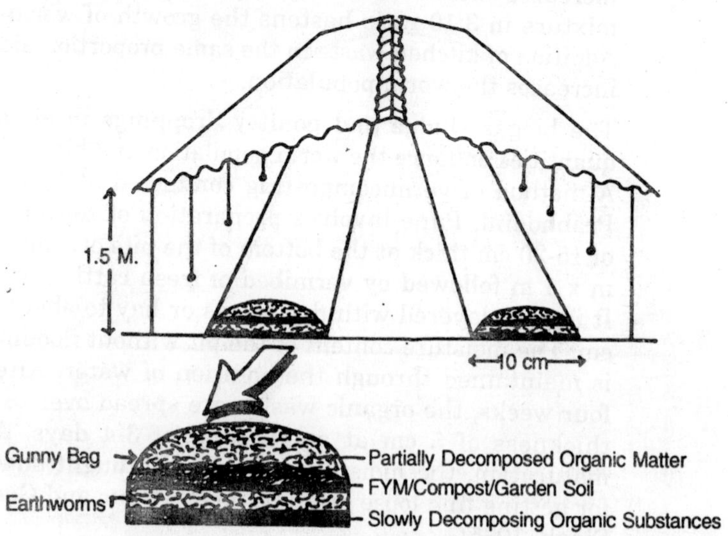

Changes During Vermicomposting

Lindwist (1941) reported that earthworm increases the nitrate production by stimulating bacterial activity and through decomposition of their own bodies. The presence of earthworm had a marked effect upon nitrogen transformation in a tissue waste or cow slurry mixture. Nitrogen mineralization was greater in the presence of earthworms and this mineral nitrogen was retained in the NO3 form which is due to the favourable condition for nitrification product by *E. foetida*. Earthworms assimilate organic nitrogen and excrete approximately equal amount of nitrogen as ammonium and muco-proteins (Needha, 1957). Feedng on aerobic sewage sludge and domestic animal manure, earthworms increase their overall rate of decomposition (Mitchell *et al.*, 1977 and Horonor and Mitchell, 1981). It has been observed that earthworms decrease the proportion of anaerobic to aerobic decomposition which results in a decrease of methane and volatile sulphur compounds (Mitchell *et al.*, 1980 and Waugh and Mitchell, 1981). Edwards (1981 and 1982) found that worm activity in

Table 6: Changes in physico-chemical properties during humification of different residues with and without earthworms (*Eisienia foetida*)

Sr. No.	*Particulars*	*pH*		*Organic carbon*		*Nitrogen (%)*		*Phosphorus (%)*		*Potassium (%)*		*Calcium (%)*	
		I	F	I	F	I	F	I	F	I	F	I	F
1.	Cow dung	7.75	7.84	35.32	41.50	1.65	2.02	0.79	1.03	0.38	0.72	0.82	1.39
2.	Cow dung + earthworms	7.76	7.41	35.28	35.61	1.68	2.27	0.82	1.13	0.42	0.83	0.85	1.69
3.	Farm waste	7.54	7.13	38.60	25.23	1.13	1.51	0.40	0.53	0.41	0.61	0.40	0.61
4.	Farm waste + earthworms	7.52	7.25	39.78	24.11	1.18	1.53	0.39	0.57	0.40	0.72	0.43	0.89
5.	Local grass	6.99	6.68	39.05	29.37	1.45	1.70	0.40	0.50	0.36	0.48	0.41	0.75
6.	Local grass + earthworms	6.87	6.70	39.61	27.43	1.48	1.79	0.43	0.59	0.39	0.60	0.39	1.08
7.	Mango leaves	6.16	6.55	43.10	37.01	1.11	1.30	0.38	0.51	0.29	0.54	0.94	1.38
8.	Mango leaves + earthworms	6.15	6.72	43.99	33.65	1.10	1.67	0.36	0.56	0.30	0.62	0.97	1.67
9.	Cashew leaves	6.13	6.26	42.79	36.07	1.07	1.34	0.37	0.50	0.28	0.42	0.51	0.81
10.	Cashew leaves + earthworms	6.14	6.49	43.28	33.52	1.09	1.43	0.37	0.53	0.28	0.50	0.54	1.09
11.	Forest leaves	7.44	6.98	41.60	35.53	1.35	1.67	0.37	0.55	0.32	0.54	0.68	1.19
12.	Forest leaves + earthworms	7.47	7.18	41.85	33.96	1.57	1.90	0.38	0.61	0.34	0.62	0.71	1.51

I = Initial, F = Final (After 150 days)

Table 6: (Contd...)

Sr. No.	Particulars	Magnesium		Copper (ppm)		Manganese (ppm)		Zinc (ppm)		Humic acid (ppm)		Fulvic acid (ppm)	
		I	F	I	F	I	F	I	F	I	F	I	F
1.	Cow dung	0.21	0.44	0.89	1.47	5.16	7.08	2.01	2.40	5.23	6.58	3.60	4.10
2.	Cow dung + earthworms	0.22	0.61	0.88	1.65	5.19	9.96	2.09	2.70	5.23	6.94	3.56	4.58
3.	Farm waste	0.14	0.37	1.69	1.89	16.62	18.72	3.09	3.48	3.30	4.28	2.38	3.58
4.	Farm waste + earthworms	0.17	0.44	1.68	2.03	16.59	20.76	3.06	3.90	3.31	4.60	2.35	3.95
5.	Local grass	0.09	0.29	1.73	1.93	5.61	8.52	1.71	1.92	4.13	4.25	3.87	4.09
6.	Local grass + earthworms	0.13	0.43	1.74	2.07	5.67	9.78	1.74	2.10	4.05	4.74	3.83	4.36
7.	Mango leaves	0.21	0.44	1.67	1.77	11.79	14.85	1.38	1.56	4.30	5.11	2.51	3.21
8.	Mango leaves + earthworms	0.26	0.57	1.69	1.88	11.85	16.20	1.35	1.74	4.15	5.37	2.46	3.51
9.	Cashew leaves	0.13	0.35	1.53	1.72	7.68	10.62	1.62	2.22	4.12	6.06	2.80	3.04
10.	Cashew leaves + earthworms	0.16	0.44	1.55	1.83	8.28	12.51	1.68	2.28	4.18	6.40	2.78	3.48
11.	Forest leaves	0.20	0.40	1.52	1.66	10.95	13.50	2.10	2.28	4.81	7.03	3.75	4.61
12.	Forest leaves + earthworms	0.21	0.49	1.52	1.74	10.98	17.25	2.04	2.82	4.77	7.48	3.70	5.19

I = Initial, F = Final (After 150 days)

the animal waste changed most of the nitrogen from ammonium to the nitrate form Riffalda and Leximinzi (1983) observed that manure samples inoculated with *E. foetida* decompose more rapidly and showed a higher degree of humification than uninoculated samples. In general, pH values towards neutrality were maintained in the treatments inoculated with earthworms as compared to non-inoculated residues. As suggested by Wallwork (1983), all enzymes are active in a very narrow pH range and earthworm bioreactors efficiently maintain the high non-linear parameter like pH. The active calciferous gland in the earthworms contain certain large quantities of carbonic anhydrase which catalyse the fixation of carbon dioxide in the form of calcium carbonate, preventing the fall in the pH of the body fluid. Albanell *et al.* (1988) compared casting of *Eisenia foetida* from sheep manure alone or mixed cotton wastes with the same manures in the absence of earthworms which were analysed every two weeks for three months. They concluded that earthworm accelerated the mineralization rate and converted the manures into castings with a higher nutritional value and degree of humification. The castings obtained from manure mixed with cotton wastes showed good fertilizing quality, suggesting that this kind of industrial residue may be used in the vermicomposting. Gaur (1992) showed that a C:N ratio of 30-40 is optimum for efficient composting. High C:N ratio is generally caused by organic material poor in nitrogen such as straw of cereals, sugarcane trash, maize stalk, cotton stalk, jute stems etc. They found that due to inoculation of microorganisms, the period of composting was reduced by one month and quality of compost was improved. The nitrogen content, available phosphorus and humus content increased in inoculated treatment over respective control.

Hand *et al.* (1988) studied the changes during composting of cow dung slurry with and without *E. foetida* for a period of 35 days. Most of the initial ammonium nitrogen contained in the slurry was lost by 14 days, probably largely due to volatilisation of ammonia. In the treatment containing earthworms, nitrate nitrogen was conserved throughout the experiment while in the slurry without worms, a small initial increase in NO3 occurred and the level of this constituent subsequently fell to a level below that of the beginning of the experiment. Organic carbon content of the mixture containing worms declined rapidly at the beginning of the experiment but stabilized after 21 days. At the end of this experiment, carbon content of the slurry was lower than that of slurry without worms. The C:N ratio of the slurry containing earthworms decreased rapidly initially as organic carbon was lost from the substrate. The ratio then stabilized and finally rose slightly due to continued loss of organic nitrogen. At the end of the experiment the control mixture had a slightly lower C/N ratio than the mixture containing earthworms showed no changes during the experiment. However, in the control mixture, some evidence of cellulose decomposition was observed over six weeks. The proportion of lignin increased slightly with time in the mixture containing earthworms, but showed such a greater increase in the control mixture. Total number of bacteria in the two mixtures were similar for six weeks, showing some trend of a slow decline between six and seven weeks, however, the number of bacteria in the mixture containing earthworms fell sharply compared with the control. The significant effect of certain microorganisms upon earthworm growth suggested that a vermicomposting system could be developed in, while substrate could be seeded with suitable microorganisms to enhance the efficiency of vermicomposting. The influence of earthworms on the decomposition of ^{14}C labelled plant material in soil was studied by Cheshire and Griffiths (1989). In their experiment, amounts of the individual carbohydrates components derived from informally ^{14}C labelled grass added to soil were monitored in incubations lasting for upto two years. Decomposition of grass was enhanced by the presence of earthworms. Earthworms have increased the loss of ^{14}C xylose by 5%, glucose by 21%, 14% and 8%, respectively after 12 months. Earthworms after 3 months, become progressively less. Total ^{14}C losses in presence of earthworms were 66% after 1 year and 60% after 2 years. Increased decomposition is considered to be the result of the mixing of soil and substrate by the invertebrates, rather than an effect of their digestive capabilities.

Jadrijevic *et al.* (1991) observed that in cattle and goat manure, *E. foetida* activity contributed to variation in C:N relationship, CO2 release and number of microorganisms and it increased bulk density and particle aggregation in both manures. Manna *et al.* (1994) studied the changes in ash content, water soluble carbohydrate, C/N ratio, cation exchange capacity and biodegradability index of various organic wastes inoculated with vermiculture after 180 days of composting. The results indicated that inoculation of the material with earthworms significantly increased the ash content and CEC of all the waste materials, but the increase in CEC waste materials, but the increase in CEC inoculation with earthworm was relatively less as compared to control. Though the C:N ratio was narrowed down in both inoculated materials yet the magnitude of effect was larger in the former case. Similarly, significant reduction in both water soluble carbohydrates and water soluble carbon was noted in both the treatments. The minimum value of biodegradability index value was recorded in the city garbage followed by maize stalk, chickpea straw, wheat straw and soybean straw.

Nutrient Value of Worm Cast and Vermicompost

The production of cast by earthworms depends on the season and the type of vegetation. It is observed that maximum worm cast production occurred in rainy season and minimal production in summer. Worm cast production in grassland is reported to be 133 tons dry weight/ha/year and about 98 tons by dry weight/ha/year in woodland areas of Karnataka. The quantitative analysis of worm casts, mucus and dead worm tissue revealed that about 230 kg N/ha/year is contributed to the soil in the grassland site and 165 kg N/ha/year in the soil of woodland site (Krishnamoorthy, 1984). The chemical composition of worm cast of *E. eugeniae* is given in Table 7. The casting contain as much as 5 times more nitrate nitrogen, 14 times more calcium, 3 time more magnesium 11 time more potassium than that of 15 cm top soil (Sharma and Madan, 1983).

There are reports that concentration of exchangeable cations such as Ca, Mg, Na, K, available P and Mo in worm casts is more than in the surrounding soil (Shinde *et al.*, 1992).

Table 7: Chemical composition of worm cast

Sr. No.	Property	Value
1.	pH	7.1
2.	Electrical conductivity (mhos/cm)	1.0
3.	Organic carbon (%)	4.6
4.	Nitrogen (%)	0.5
5.	Phosphorus as P2O5 (%)	1.2
6.	Potassium as K2O (%)	0.1

(Talashilkar, 1986)

On the whole, vermicompost can not be described as being nutritionally superior to other organic manures but unique way in which it is produced, even right in the field and at low cost makes it very attractive for practical application. The methods followed by different workers vary a great deal and steps taken are sometimes arbitrary resulting variation in product quality. Therefore, there is a need to standardise the method of vermicomposting for obtaining uniformly good quality products. Bhole (1992) and Deshpande (1992) have also noted similar observations. Shrikhande and Pathak (1951) reported that earthworm casting increases the availability of nutrients to greater extent than the termite gallaries and hill and control. Carbon to nitrogen ratio has been reported by several workers to be higher in worm cast than in the soil in which they live (Satchell, 1963 and Graff, 1971). Ghabbour (1966) stated that besides microorganisms, inorganic minerals and organic matter, the cast also contains enzymes such as proteases, amylases, lipases, cellulases and chitinases which continue to disintegrate organic matter even after they have been execrated. Sharma and Madan (1983) reported that earthworm casting contain as much as 5 times more nitrate nitrogen, 14 times more calcium, 3 times more magnesium, 11 times more potassium than that of 15 cm top soil. Krishnamoorthy and Vajranabhaiah (1984) noticed that *Pheretima elongata* released 83 nanograms cytokinin per day per se and 419 nanograms auxin per day per se, which are present in intestine of worms.

Shinde *et al.* (1992) reported that vermicompost contains more carbon and phosphorus than FYM, it had less K and micronutrients than FYM and both had comparable contents of nitrogen. Vermicompost generally had wide C:N ratio as compared to FYM.

Kale (1998) reported that the biodegradable organic waste can be converted into vermicompost. When earthworms feed on organic wastes it undergoes physical and chemical breakdown during processes of ingestion and digestion. About 5-10 per cent of the ingested material is absorbed into the tissue for their growth and metabolic activity and rest is excreted as cast. The cast is mixed with mucus secretion of gutwall and of the microbes. These add to structural stability of the cast which is used as vermicompost. The decomposition process continues even after the release of the cast by the establishment of microorganisms. The levels of nutrients in the vermicompost will be at lower level than in the original material as in compost derived from any other method. But levels of macro and micronutrients will be higher than in the compost derived from alternative methods. The nutrient level depend upon nature of organic waste used as food source of earthworms. Nutrient status of vermicompost on using different organic waste as food source is as follow: organic carbon, 9.15 to 17.98%; total nitrogen, 0.5 to 1.5%; calcium and magnesium, 22.70 to 70 mg/100 gm; copper 2 to 9.5 ppm; iron, 2 to 9.3 ppm; zinc, 5.7 to 11.5 ppm and available sulphur, 128 to 548 ppm.

Reddy and Reddy (1998) conducted an experiment for two consecutive years during kharif and rabi seasons of 1994-95 and 1995-96 at Hyderabad on sandy loam soil taking maize (DHM-105) in kharif and soybean (Hardee) in rabi season with four levels i.e. 0, 25, 75 and 100 per cent substitution of recommended dose of nitrogen by vermicompost, poultry manure, biogas slurry and FYM. They have analysed nutrient content of organic manures and given chemical composition of vermicompost. Vermicompost contains 1.98% N, 1.23% P and 1.59% K, total and DTPA extractable micronutrients status were also given as total Zn 132.0 mg kg^{-1}, Cu 70.5 mg kg^{-1}, Fe 1440.2 mg kg^{-1} and Mn 317.5 mg kg^{-1}, DTPA extractable

Zn 62.0 mg kg^{-1}, Cu 15.8 mg kg^{-1}, Fe 321.6 mg kg^{-1} and Mn 79.2 mg kg^{-1}. In order to identify the best organic waste for vermicomposting an experiment was conducted by Kachhave and Jaishankar (1999) with saw dust, city waste, sugarcane trash, weed plant (*Parthenium* spp.), pressmud and slaughter house waste under two methods of composting, pit and pot methods. The total carbon content of vermicompost from all organic waste were appreciably enhanced. Total carbon content of pressmud vermicompost recorded increase of 10.2 per cent and 10.9 per cent over control in pit and pot cultures, respectively. The highest enhancement of 1.22 per cent N was recorded for slaughter house waste and least with cane trash. The maximum level of P and K were recorded for pressmud vermicompost and least enhancement for saw dust vermicompost. The wider C:N ratio was obtained for vermicompost prepared from cane trash while narrowest C:N ratio were found to slaughter house waste followed by city waste vermicompost.

To evaluate efficacy of vermicompost prepared from different organic materials like sugarcane trash, ipomea weed, neem leaves, parthenium weed and banana waste, Vasanthi and Kumaraswamy (1999) conducted field experiment during 1994-95 and 1995-96 on red sandy clay loam soil at College of Agriculture, Madurai. The data on nutrient status showed that contents of macro and micronutrients were more by many times in composted materials compared to their contents in raw material. Desirable reduction in C:N ratio was observed in enriched compost manure. Nitrogen content in vermicompost prepared from organic waste followed the following order: ipomea weed (2.99%) > banana waste (2.83%) > parthenium weed (2.99%) > sugarcane trash (2.67%) > neem leaves (2.61%). Similarly, percentage phosphorus content in vermicompost prepared from organic waste followed the order: ipomea weed (1.37%) > parthenium weed (1.20%) > banana waste (1.18%) > neem leaves (1.17%) > sugarcane trash (1.06%). The potassium content of vermicompost followed the order: ipomea weed (1.46%) > banana waste (1.32%) > parthenium weed (1.19%) > neem leaves in Ca, Mg, Fe, Mn, Zn and Cu content of vermicompost as compared to raw organic waste.

Effect of Vermicompost on Mineralization of Nitrogen, Phosphorus and Potassium from Soil

During the process of decomposition of organic materials in soil, organic acids formed in the transformation cycle and also those excreted by soil orgnisms can contribute to the solubilization of nutrients from mineral components of soil or parent material. The organic matter is the most effective cation exchanger and thus stores the nutrients against leaching. Further active biodegradtion and resultant production of CO2 can lead to dissolution of nutrient in soils. The triggering of oxidation – reduction reactions, particularly in flooded soils and increased chelation capacity brought about by addition of organic matter dictate transformation and availability of nutrients in soil (Waksman, 1948 and Sarkar *et al.*, 1998). Sicar *et al.* (1940) noted that the mineralization of nitrogen occurred at much higher C/N ratio under anaerobic condition than under upland or aerobic conditions. This is attributed to the low nitrogen requirement of anaerobes mainly bacteria. Frapes and Sterges (1947) have observed that the nitrogen mineralization is increased due to higher content of nitrogen in soil. Jannson *et al.* (1955) observed an initial liberation of substantial quantity of nitrogen followed by decreasing lower rate of mineralization as decomposition of organic matter progressed. Mitsui (1955) stated that air drying the soil prior to flooding almost doubled the quantity of ammonia released.

Dodalto and Costa (1990) showed that cast of *Glossoscolex* spp. had a higher content of available nutrients, particularly available phosphorus than the adjoining soil. Jambhekar (1990) reported that application of vermicompost increased the available N, P and K content in soil. Results of work done at Ganeshkhind revealed that there was decrease in pH due to application of FYM and vermicompost. However, organic carbon, available N, P and K were increased due to FYM and vermicompost application as compared with control (Anonymous, 1992).

Reddy and Reddy (1998) conducted experiment for two consecutive years during kharif and rabi season of 1994-95 and 1995-96 using combinations of four levels of 25, 50, 75 and

100 per cent substitution of recommended dose of nitrogen by vermicompost, poultry manure, biogas slurry and FYM. In all types of manures the treatment with 100 per cent level of vermicompost which was on par with 75 and 50 per cent level of manures showed highest available nitrogen content i.e. 220.20 kg ha^{-1} over the treatments receiving poultry manure, biogas slurry, FYM and control treatments. They also found that full dose (100% N) of vermicompost showed higher content of P2O5 (38.67 kg ha^{-1}) over treatments control and recommended dose of fertilizers i.e. 13.15 and 30.29 kg ha^{-1}, respectively after two cropping cycles under maize-soybean cropping system. The increase might be due to release of organic acids during microbial decomposition of organic matter. Organic matter leads to formation of coating on sesquioxides which reduces phosphate fixing capacity of soil. The 100% substitution of recommended dose of nitrogen by vermicompost showed highest content of potassium (318.95 kg ha^{-1}) over the treatments control (223.85 kg ha^{-1}) and recommended dose of fertilizers (308.35 kg ha^{-1}) after two cropping cycles. This may be due to reduction of potassium fixation and organic matter interaction with K clays to release K from non exchangeable fraction to the available pool.

Vasanthi and Kumaraswamy (1999) conducted field experiments during 1994-95 and 1995-96 on red sandy loam soil at Madurai to evaluate effect of vermicompost prepared from different organic materials to increase the soil fertility status. The pooled results of three experiments showed that available nitrogen status in the soil was significantly higher in the treatments that received vermicompost besides N, P and K. The available nitrogen status ranged between 192 kg ha^{-1} in NPK alone treatment and 290 kg ha^{-1} in ipomea vermicompost plus N, P and K treatment. The average available nitrogen status in the soil at 5 and 10 t ha^{-1} of vermicompost was 242 and 239 kg ha^{-1}, respectively. The results also showed that labile phosphorus status ranges between 11.4 kg ha^{-1} in N, P and K alone treatment (100:50:50) and 16.25 kg ha^{-1} in ipomea vermicompost plus NPK treatment. The average labile phosphorus status at 5 t and 10 t ha^{-1} of vermicompost was 12.41 and 13.61 kg ha^{-1}, respectively. The results showed that significant positive increase in labile phosphorus content in the treatments that received one of the vermicompost prepared from ipomea weed, parthenium weed, neem leaves, cane trash, banana waste separately. The pooled results of three experiments showed that available potassium status ranged between 225 kg ha^{-1} in N, P and K alone treatment and 458 kg ha^{-1} in cane trash vermicompost with NPK treatment. The average available potassium status at 5 and 10 t ha^{-1} levels of vermicompost was 284 and 371 kg ha^{-1}, which showed that application of vermicompost @ 10 t ha^{-1} recorded significant increase in available potassium status of the soil.

Hapse *et al.* (1993) observed that application of vermicompost @ 5 t ha^{-1} significantly increased total N, available N, P and K and organic carbon and decreased pH over control. Application of vermicompost also decreased the bulk density and increased the porosity of soil as compared to control. Rao (1994) studied the effect of compost and vermicompost on soil properties and biomass production in maize. The vermicompost had narrower C:N ratio compared to compost. Vermicompost was found to be superior over compost in releasing available nutrients in soil since it contained higher amount of N, P, K, Ca, Mg, S and micronutrients as compared to compost. Jadhav (1995) reported that use of earthworms had significantly increased the per cent phosphorus content over uninoculated organic residues, which may be due to contribution of phosphorus by worm cast as reported by Shinde *et al.* (1992). Patil (1993) studied the effect of application of vermicompost and FYM on release of nutrients and their uptake and yield by maize in different textured soil. Their studies revealed that application of FYM and vermicompost resulted into significant increase in electrical conductivity, organic carbon, available N, P and K contents of all the soil types, whereas pH of all soil types significantly decreased. The availability of N, P and K was observed more in clay soil than the clay loam and sandy soils. Application of vermicompost significantly improved the physical properties of all the soil types under study. Vermicompost application showed the significant improvement in chemical properties of soil. Patil *et al.* (2000) used four organic sources *viz.*, spent wash, press mud compost, vermicompost, press mud and FYM conjointly with potassium to study their effect on soil properties and yield of wheat. Application of spent wash and press mud compost @ 15 t/ha increased soil EC, organic carbon and available N, P and K status of soil. Press mud increased available P status of soil and graded levels of potash increased the soil available N, P and K. Spent wash press mud compost and 60 kg K2O ha^{-1} increased N, P and K content of soil.

Inamdar (2001) studied the effect of vermicompost and compost applied at the rate of 1%, 2% and 4% organic carbon basis on the mineralization of nutrients and changes in humus fractions in lateritic and medium black soils of Konkan region of Maharashtra. He concluded that the nutrient availability from native as well as applied sources of organic manures *viz.*, vermicompost and ordinary compost increased gradually as a function of period of incubation under submergence. Increasing doses of organic manures from 1 to 4 per cent organic carbon basis resulted in corresponding increase in availability of N, P and K and rate of humification of organic manures. The availability of major nutrients was faster in vermicompost treated soil as compared to ordinary compost treated soil indicating superiority of vermicompost (Table 8, Table 9).

Effect of Vermicompost on Soil Fertility

Use of vermicompost promotes soil aggregation and stabilizes soil structure. This improves the air water relationships of soil, thus increasing the water retention capacity and encouraging extensive development of root system of plants. The mineralization of nutrients is observed to be enhanced resulting into boosting up of crop productivity (Lunt and Jacobson 1944; Graff 1971 and Satchell 1983). Russel (1909) reported that earthworms decomposed organic matter quickly and increased nitrification, which became responsible for increasing crop yield. There is much more nitrate nitrogen in worm casts than in the parent soil, moreover, the casts have a higher base exchange capacity and more exchangeable calcium, magnesium, potassium and available phosphorus than

Table 8: Changes in available nitrogen content (kg ha^{-1}) of lateritic soil

Treatments	*Incubation period (days)*													*Treatment effect*
	1	**7**	**14**	**28**	**42**	**56**	**70**	**84**	**98**	**112**	**126**	**140**	**154**	
Control	228	214	246	224	230	266	283	293	268	306	402	287	237	271
Vermicompost @ 1 % organic carbon	252	243	404	457	442	500	461	445	325	305	514	397	309	395
Vermicompost @ 2 % organic carbon	278	262	473	456	433	494	466	416	391	433	599	470	416	424
Vermicompost @ 4 % organic carbon	319	372	505	492	533	631	566	572	581	656	700	508	457	530
Compost @ 1 % organic carbon	218	274	271	278	334	429	356	315	293	377	410	426	341	332
Compost @ 2 % organic carbon	268	315	306	304	356	486	454	391	410	473	482	489	356	392
Compost @ 4 % organic carbon	416	385	420	420	461	666	500	533	549	659	606	590	461	513
Period effect	275	295	375	378	398	482	441	424	402	470	536	452	368	

	Treatment effect	*Period effect*	*Interaction effect*
S.E. ±	3.2	4.4	11.9
C.D. (at 5 %)	9.1	12.4	33.0

Table 9: Changes in per cent humic acid content of lateritic soil

Treatments	*Incubation period (days)*													*Treatment effect*
	1	**7**	**14**	**28**	**42**	**56**	**70**	**84**	**98**	**112**	**126**	**140**	**154**	
Control	0.31	0.40	0.48	0.78	1.26	1.46	1.48	1.47	1.49	1.52	1.61	1.60	1.57	1.17
Vermicompost @ 1 % organic carbon	0.80	0.53	0.51	0.89	1.48	1.55	1.61	1.70	1.72	1.63	1.67	1.65	1.64	1.33
Vermicompost @ 2 % organic carbon	1.21	0.83	0.78	1.02	2.50	3.20	3.40	3.39	3.35	3.39	3.21	3.21	2.83	2.49
Vermicompost @ 4 % organic carbon	3.0	2.86	2.60	3.10	4.63	5.10	5.40	5.52	5.49	5.58	5.49	5.46	5.40	4.58
Compost @ 1 % organic carbon	0.50	0.45	0.42	0.80	1.37	1.46	1.51	1.56	1.59	1.60	1.58	1.61	1.61	1.23
Compost @ 2 % organic carbon	0.74	0.60	0.59	0.99	2.50	2.28	2.40	2.39	2.37	2.39	2.24	2.20	2.21	1.83
Compost @ 4 % organic carbon	3.02	2.95	2.83	3.11	4.32	4.93	5.08	5.13	5.14	5.22	5.19	5.16	5.15	4.40
Period effect	1.36	1.23	1.15	1.52	2.58	2.86	2.98	3.02	3.02	3.04	2.99	2.98	2.91	

	Treatment effect	*Period effect*	*Interaction effect*
S.E. ±	0.01	0.007	0.03
C.D. (at 5 %)	0.03	0.020	0.08

the soil in which the worms live (Lunt and Jacobson, 1944; Nijhawan and Kanwar, 1952b and Graff, 1971) Lindwist (1941) reported that the earthworms increased the nitrate production by stimulating bacterial activity and through decomposition of their own bodies. Lunt and Jacobson (1944) and Graff (1971) stated that the amounts of available mineral nutrients are also improved in soil containing earthworms and it is generally concluded that such soils have higher concentration of exchangeable Ca, Mg, K and available P. Nijhawan and Kanwar (1952a) observed that application of earthworm compost to wheat increased the availability of N, P and K and decreased the pH of the soil as compared to control. Stockdill and Cossens (1966) reported the influence of worm activity and worm cast application in field on the improved growth of pastures.

Gupta and Sakal (1967) in their comparative study of cultivated and garden soil casts observed that earthworm cast contained more available P, organic matter, total nitrogen and nitrate nitrogen. The C/N ratio was narrowed. Garden soil casts were rich in organic matter and nitrogen as compared to the cultivated soil castings. Cultivated soil castings were rich in nitrate nitrogen than the garden soil castings. Lakhani and Satchell (1970) reported that earthworms can process about 10-25 per cent of the net annual energy input into the system which indicates their place of improtance in the decomposer system and in improving soil fertility. They reported that certain metabolites produced by earthworms may be responsible for stimualting plant growth (Gavrilov, 1962 and Nielson, 1965). It is believed that earthworms release certain vitamins and similar substances into the soil which may be the B group vitamins (Gavrilov, 1963) or some pro-vitamin D (Zrazhevskii, 1957) or free amino acids (Dubash and Ganti, 1964). The amount of vitamin B is generally known to double itself twice within a year or two after the introduction of these worms Atlavinyte *et al.* (1971). Edwards and Lafty (1972) inferred that earthworm can also improve the amount of available molybdenum in soil. Haini and Hunta (1987) studied the physical and chemical analysis and observed only minor differences between worm worked and wormless wastes; thus vermicompost could be considered superior to ordinary compost only with regard to its physical structure. Dadalto and Costa (1990) showed that cast of *Glossoscolex* spp. had a higher content of available nutrients, particularly available phosphorus than the adjoining soil. Jambhekar (1990) reported that application of vermicompost increased the available N, P and K content in soil. Results of work done at Ganeshkhind revealed that there was decrease in pH due to application of FYM and vermicompost. However, organic carbon, available N, P and K were increased due to FYM and vermicompost application as compared with control (Anonymous, 1992). Kale *et al.* (1992) studied the influence of vermicompost application on the available macro nutrients and selected microbial populations in a rice field. They found that the vermicompost application has enhanced the activity of these selected microbes in the soil system. There was high level of total N in the experimental plot which comparatively received less quantity of fertilizers. This may be due to more N-fixers observed in the experimental plot (3.48×10^3) than that of the control plot (2.16×10^3). They have reported that concentration of exchangeable cations such as Ca, Mg, Na, K, avaialble P and Mo in wormcasts is more than in surrounding soil (Shinde *et al.*, 1992) on the whole vermicompost can not be described as being nutritionally superior to other organic manures but unique way in which it is produced even right in the field and at low costs makes it very attractive for practical application.

Hapse *et al.* (1993) observed that application of vermicompost @ 5 t/ha significantly increased total N, available N, P, K and organic carbon and decreased pH over control. Application of vermicompost also decreased the bulk density and increased the porosity of the soil as compared to control. Shinde and Gawade (1992) and Patil (1993) studied the effect of application of vermicompost and FYM on release of nutrients and their uptake and yield by maize in different textured soils. Their studies revealed that application of FYM and vermicompost resulted into significant increase in electrical conductivity, organic carbon, available N, P and K content of all the soil types, whereas the pH of all the soil types significantly decreased. The DTPA extractable micro nutrients *viz.*, Fe, Mn, Cu and Zn content of soil was significantly increased upto 30 days in incubation and further decreased in all the soil types. The availability of N, P and K was observed more in clay soil than in clay loam and sandy soils. Application of vermicompost significantly improved the physical properties of all the soil types under study. Vermicompost application showed the significant improvement in chemical properties of soil. Rao (1994) studied the effect of compost and vermicompost on soil properties and biomass production in maize. The vermicompost had the narrower C:N ratio compared to compost. Vermicompost was found to be superior to compost in releasing available nutrients in soil, since it contained higher amounts of N, P, K, Ca, Mg, S and micronutrients when compared to compost. Addition of vermicompost and compost in combination with the fertilizers had increased the soil pH, while the application of fertilizers alone had decreased the pH Jambhekar (1994) also noted beneficial effects of vermicompost on available potassium status of soil in grape garden. Jadhav *et al.* (1997) also observed higher values of electrical conductivity of the soil treated with vermicompost in comparison to the treatments without vermicompost. The reasons ascribed to this are increased retention of exchangeable bases and enhanced buffering capacity of the soil arising out of the addition of organic manures. Nethra *et al.* (1999) reported that the continuous use of fertilizers is found to have detremental effects on the soil health and environment. Since there is increased focus towards organic farming round the world an attempt was made to study the effect of vermicompost (manure obtained after subjecting organic waste to earthworm activity) as an organic amendment in growing China aster (*Callistephus chinensis* L Nees.). It was observed that the maximum plant height (33.33 cm), number of leaves (33.08) and yield/ha (6840 kg/ha) were recorded after application of 10 t/ha vermicompost and recommended N,P,K. Vasanthi and Kumarswami (1999) studied the efficacy of vermicompost to improve soil fertility after the harvest of three crops of rice. The results of the experiment indicated that organic carbon content and fertility status as reflected by the available status of N, P, K and micronutrients and CEC were higher and bulk density lower in the treatments that received vermicompost plus N, P and K than in the treatment with N, P and K alone.

Vermicompost and Crop Production

Earthworm enhanced plant growth has usually been attributed to improved soil physical and chemical conditions but a direct harmone like effect on plant has been demonstrated by Nielson (1951 and 1965). Nijhawan and Kanwar (1952a) observed that application of earthworm compost to wheat increased the plant height, number of tillers, number of leaves, early earheading, earhead length and dry matter per plant than control. Dash (1978) suggested that worms alongwith organic manures can be utilized as an alternative to costly inorganic fertilizers for growing crops. Tomati *et al.* (1983) reported that earthworm cast which is rich in available nutrients increased the plant growth and yield of crop. Though Kale and Bano (1984) reported that differences in grain yield of rice between the treatment plots receiving vermicompost @ 1.5 t/ha and inorganic fertilizers such as urea @ 100 kg N/ha plus Suphala (15:15:15) @ 200 kg/ha were statistically at par. Grappelli *et al.* (1985) worked on rooting experiments on layers of *Ficus elastica, Diffenbachia amoena, Dracaena deremensis* and on shoots of *Ficus elastica, Begionia, Massohiana, Aglashoema costatum* were performed to ascertain the possibility of employing earthworm casting in plant propagation. Racking of layers was favoured when casting alone was supplied whereas rooting of shoots generally was enhanced when casting mixtures were used. Kale *et al.* (1987) found that worm cast when used as manure increased height of plant, leaf area index, number of branches, stem girth and yield in respect of plants like salvia and aster in pots. Sharma and Madan (1988) studied the effects of various organic wastes such as garbage, compost, turning matter, screening matter, poultry wastes, sawdust, biogas slurry, cattle dung and sewage sludge alone and with earthworms on the total dry matter yield of wheat and maize. The results showed that total dry matter yield of wheat and maize increased significantly with the application of various organic wastes @ 1 per cent with earthworms over the treatment receiving various organic wastes alone except sawdust. Jambhekar (1990) reported that application of vermicompost increased the available N, P and K content in soil. There was a significant increase in the uptake of N, P and K by grape. Tomati *et al.* (1990) reported increase in protein synthesis in radish, when grown in the presence of worm cast.

Results of work done at Ganeshkhind revealed decrease in pH due to the application of vermicompost. However, organic carbon, available N, P and K were increased due to vermicompost application as compared with control (Anonymous, 1992). Patil (1993) studied the effect of application of vermicompost as compared to control. N, P and K content in maize and their uptake by maize at tasseling and harvesting stages were significantly increased with the vermicompost application. Kang *et al.* (1994) studied the effect of worm casts prepared from agroforest woody spice, *Dactyladenia barteri, Glyricidia sepium, Leuceaena leucocephala, Senna siamea and Treculia africana*. Despite the high nutrient level in the worm casts, nutrient availability for crops is observed to be low. Worm casts thus, appeared to serve as a slow release nutrient source in the soil. There were significant differences in N, P and K uptake by maize in different soils. Ranganathan and Christopher (1994) reported that the vermicompost not only helped to protect fertility of the soil but also boosted productivity depending on crops, season and other factors and also enhanced the quality of end products. The continual use of vermicompost over the year has also resulted in reduction of pests and disease problems. The response was most pronounced in grapes followed by pomegranate, banana, papaya, guava, sugarcane, cereals and other crops. Palanisamy (1996) studied the effect of mixing 10 to 20 per cent worm cast in the upper 30 cm of a sandy soil on the growth parameters and yield of wheat. He observed 40 per cent increase in yield of wheat due to mixing of cast in 30 cm soil. Similarly, an increase in shoot length and total dry matter were also noted. When soyabean seeds cv. CO-1 were pelleted with vermicompost @ 50 g/kg seed, the pod yield was increased by 16 per cent (Dharmalingam, 1996).

Bawa (1995) studied an efficacy of vermicompost in comparison to other organic manures such as FYM, city compost and glyricidia to groundnut crop. He found that the yields of groundnut were at par in all the treatments receiving said organic manure @ 10 t/ha. An application of vermicompost @ 10 t/ha alongwith recommended dose of N and P resulted into 55 per cent increase in dry pod yield over the treatment receiving recommended dose of N and P (Table 10). In another experiment, an application of vermicompost @ 10 t/ha + 50 per cent recommended dose of NPK produced 63.06 q/ha of rice yield which was significantly higher over the treatment

Table 10: Effect of vermicompost and other organic manures on the pod yield of groundnut

Sr. No.	*Treatments*	*Pod yield (q/ha)*
1.	Control	14.36
2.	25 kg N + 50 kg P2O5/ha	17.89
3.	FYM @ 10 t/ha	17.19
4.	N + P + FYM @ 10 t/ha	26.88
5.	City compost @ 10 t/ha	17.47
6.	N + P + city compost @ 10 t/ha	27.04
7.	Vermicompost @ 10 t/ha	17.36
8.	N + P + vermicompost @ 10 t/ha	26.92
9.	Glyricidia @ 10 t/ha	27.43
10.	N + P + glyricidia @ 10 t/ha	27.11
	S.Em. ±	0.31
	C.D. at 5%	0.89

Bawa (1995)

receiving recommended dose of NPK (52.22 q/ha) (Anonymous, 1996) (Table 11). Jadhav (1995) studied the effect of FYM and vermicompost on the yield and uptake of nutrients by rice and on the physico-chemical properties of lateritic soil. He concluded that when vermicompost was used as a source of organic manure for substitution of Urea-N to rice, it is possible to reduce 50 per cent of the recommended dose of Urea-N by vermicompost. An application of vermicompost also resulted into significant increase in nutrient status of soil (Table 12). Ghosh *et al.* (1996) studied use of vermicompost in integrated

Table 11: Grain yield of rice as influenced by vermicompost and fertilizers

Sr. No.	*Treatments*	*Grain yield (q/ha)*
1.	Recommended dose of NPK (100:50:50)	52.22
2.	Vermicompost @ 10 t/ha	47.61
3.	FYM @ 10 t/ha	45.70
4.	Glyricidia @ 10 t/ha	45.82
5.	Vermicompost @ 10 t/ha + 50% recommended dose of NPK	63.06
6.	FYM @ 10 t/ha + 50% recommended dose of NPK	62.71
7.	Glyricidia @ 10 t/ha + 50% recommended dose of NPK	60.90
	S.Em. ±	3.15
	C.D. at 5%	9.37

Anonymous (1996)

plant nutrient management and noted its additive effect on rice yields. Madhukeshwar *et al.* (1996) conducted an experiment to show the response of tomato seeds and seedlings to different nutrient substrates. The experiment was conducted by using the peat moss. Neo peat, vermicompost: sand (1:3), vermicompost: Sand (1:4), Control sand. He showed that the average leaf area (cm^2) was increased from 17 to 58 sq. cm. in the vermicompost: sand (1:3) in comparison to control sand media and the average number of leaves increased from 2 to 4 in the vermicompost: sand (1:3) in comparison to control sand. Vadiraj *et al.* (1996) studied the effect of vermicompost on cardamom seedlings in different growth media containing composting of normal mixture (Soil: Sand: FYM), FYM, compost, vermicompost, soil rite and spagnum moss. As compared to normal soil: sand: FYM mixture application of vermicompost increased the height of seedlings from 9.9 to 13.65 cm, fresh weight of plant from 2.07 to 2.97 g., dry weight of plant from 0.42 to 0.639 and number of leaves per seedling from 5 to 6 leaves.

Table 12: Effect of different combinations of urea with FYM/vermicompost on yield and uptake of nutrient by rice

Treatments	*Grain yield*	*Straw yield*	*Total nutrient uptake (mg/plot)*				
Symbol	*(g)*	*(g)*	*N*	*P*	*K*	*Ca*	*Mg*
U0	8.12	10.99	157.55	33.09	143.81	35.13	25.34
U100	18.55	22.64	376.01	80.31	475.68	97.60	72.94
U75F25	17.67	22.14	356.84	87.42	416.73	81.35	62.49
U50F50	15.01	21.32	337.33	74.02	498.70	104.87	84.99
U25F75	14.37	20.40	331.31	71.14	495.15	115.85	90.85
U0F100	10.37	18.65	373.08	50.96	371.25	63.83	50.82
U75V25	19.18	23.33	417.16	80.40	557.28	123.60	93.22
U50V50	16.04	22.84	352.84	80.70	503.61	106.10	82.26
U25V75	14.43	21.70	317.17	75.15	459.36	94.58	75.08
U0V100	11.79	20.12	244.82	59.33	334.03	63.32	48.21
C.D. (P=0.05)	2.75	1.99	35.93	8.50	63.43	10.11	6.70

*U, F and V stand for urea, FYM and vermicompost respectively; the subscripts indicate kg N/ha supplied through U, F and V as the case may be.

Jadhav (1995)

Mahendran and Kumar (1997) studied the influence of NPK application with or without biofertilizers and organic manures such as digested organic supplement (DORS) and vermicompost (VC) in cabbage cv. Hero at Kodaikanal. The highest TSS and ascorbic acid contents were produced with Dors and vermicompost. The number of non-wrapper leaves was highest in the control plots. The polar and equatorial diameters of cabbage heads and net weight were also significantly influenced by applying organic manures. Patil *et al.* (1997) investigated the effect of vermicompost, FYM and inorganic fertilizers on growth, yield and uptake of nutrients in potato cv. Kufri Chandramukhi. They recorded maximum yield of potato with vermicompost @ 4 t/ha + 50 per cent recommended dose of fertilizers (34 t/ha), net monetary income (Rs. 54, 152/ha) and cost benefit ratio (1:3.42). Thus, vermicompost @ 4 t/ha could substitute 50 per cent recommended dose of fertilizers and produce 53.46 per cent more yield over recommended dose of fertilizers. INORA (1998) conducted studies on effect of graded doses of vermicompost on crops like cotton (cv. Anjali) at MPKV, Rahuri, on sugarcane (cv. COC 671) at Kolhapur, Ugar and Ainapur (in Belgaum district) and on paddy in Pune district on cultivators fields. The results of experiment for a crop like cotton revealed that use of chemical fertilizers was observed to be very essential to achieve higher yields (Table 13), but the application of vermicompost along with chemical fertilizers (T7) increased the seed cotton yield by 6.5 per cent over the recommended dose of chemical fertilizers. Vermicompost alone (T3) increased the yield of seed cotton by 12.5 per cent over T1 (i.e. no fertilizers). Increasing the dose of vermicompost to 10 t/ha increased the yield level by 48 per cent over T1 and by 28 per cent over T3. At Ugar, application of 2.5 tonnes of vermicompost alongwith half the recommended dose of chemical fertilizers to the sugarcane crop gave the same yield (57.7 tonnes/acre) (T5) as compared to the yield obtained through recommended dose of chemical fertilizers (56.0 tonnes/acre) (T2), while at Ainapur, the application of vermicompost at 5 tonnes/ha to an Adsali crop of sugarcane gave higher cane yield of 89.4 t/acre as against 81.6 t/acre in a treatment receiving only recommended dose of chemical fertilizers (Anonymous, 1998b).

Table 13: Seed cotton yield in kg/ha

Sr. No.	*Treatment*	*Seed cotton yield (kg/ha)*	*Per cent increase in yield*
1.	Control	760.5	–
2.	Only recommended dose of C.F.	2050.6	170 over control
3.	Vermicompost alone	857.5	12.5 over control
4.	¾th VC & ¼th chemical fertilizer	1344.4	78 over control 57% over T3
5.	½ VC and ½ C.F.	1587.9	9 over 2+3
6.	¼ VC + ¾ C.F.	1940.4	150 over T1 & 126 over T3
7.	V.C. + R.D. C.F.	2181.8	6.5 over T2, 154 over T3
8.	V.C. 10 t/ha	1105.3	48 over T1, 28 over T3

VC – Vermicompost CF – Chemical Fertilizer

(INORA, 1998)

Devi Dayal and Agarwal (1998) conducted a field experiment during the spring season of 1995 ad 1996 on sandy loam soil in Hissar, Hariyana, India, Sunflowers Cv. Badshah and MSFH-8 were given 0 or 10 t farm yard manure/ha 5 or 10 t vermicompost per hectare or 250 kg/ha of fertonic (a bio-organic soil enricher) and 0, 40:20, 80:40 or 120:60 kg NP/ha. Dry matter accumulation was significantly higher in MSFH-8, while seed yield was significantly higher in Badshah. Among the organic fertilizers, seed yield was greatest with 10 t vermicompost, followed by fertonic and 5 t vermicompost, while seed yield increased with increasing NP rate. Seed yields were higher with the 2 higher rates of NP than with any of organic fertilizers. Vermicompost at 10 t/ha + 80:40 kg NP/h gave the best dry matter and seed yields. Net returns were higher with Badshah than MSFH-8. Devi Dayal *et al.* (1998) studied the comparative effect of organic manures and N + P205 rates alone and in combination on sunflowers (*Helianthus annus*) in a field experiment on sandy loam soil at Hissar during the spring seasons of 1995 and 1996. Sunflower hybrid Badshah gave a higher yield with a combination of organic manures + 80 kg N + 40 kg P205 per hectare. Maximum agronomic efficiency was recorded in hybrids Badshah and MSFH-8 with 40 kg N + 20 kg P205 per hectare application irrespective of source of organic manure, though it was highest with 10 t vermicompost per hectare. Joshi (1998) reported in an experiment in polypots that a mixture of soil. Vermicompost in ratio of 5:1 is enough to increase the growth of seedlings of 3-4 months. Further, increase in quantity of vermicompost in soil to a ratio of 4:2 and 3:3, 4:2 and 3:3 proved statistically same which indicated that the ratio between soil and vermicompost 4:2 is best for the seedlings in polybags.

Kale (1998) suggested 10 kg N + 20 kg P + 10 kg K in combination with 2 tonnes of vermicompost for obtaining optimum yields of C-152 variety of cowpea. She also reported that more than the regular macronutrients, vermicompost contributed to the supply of micronutrients essential for the crops. Apart form this, the stimulatory effect of vermicompost for nutrient uptake, growth and yield of crops is linked to the secretions of earthworms and the associated microbes mixed with the cast. Some of the organic acids that are isolated from the body fluid of earthworms and their cast have exhibited similar response as that of plant growth promoter substances. The positive responses observed in the per cent seed germination, rooting of the cuttings, hardening of the tissue cultured plants, established the view that certain organic acids in the vermicompost were the important components than the regular nutrients. Large number of cultivars of turmeric showed a positive response to yield recovery on vermicompost application at 10 t/ha although there existed a pronounced variation in degree of response. When different levels of vermicompost were tested (5, 10, 15, 20 and 25 t/ha) on another important spice coriander to quantify the vermicompost requirement, it was found that herbage and grain were maximum in all three tested varieties receiving vermicompost @ 20 t/ha. Increasing the supply of 25 t/ha did not improve herbage and grain yields further showing the need for quantifying the organic manures to get the expected yields.

When vermicompost was replaced by regular compost, expected yield was recorded on reducing the chemical fertilizers to 50 per cent. On application of vermicompost at 12 t/ha alongwith the recommended dose of chemicals, the yield recovery of bhendi (*Abelmoschus esculentus*) was 109.5 per cent more than that of the yield data recorded in package of practices irrespective of seasonal variations. In case of cowpea and bitter gourd, reducing the chemical fertilizes to 50 per cent of the recommended doses alongwith vermicompost had no adverse effect on the yield. When the nitrogen requirement was met completely through vermicompost, the yield recovery was on par with treatments receiving the chemical fertilizers (Kale,1998). A number of field and laboratory experiments were conducted by Kale (1998) and result of the same are summarised as follows: the cuttings planted in the vermicompost or exposed to slurry of vermicompost for definite time interval induced the rooting in these crops. In Mulberry cuttings, the results were effective as that of exposure to 50 ppm IBA. From the practical view point, it was felt that vermicompost production will be a low cost technology for nurseries to raise the seedling.

In pot experiments, the influence of different forms of nitrogen such as FYM, vermicompost produced from waste activated sludge and waste from a meat processing factory and ammonium nitrate on yield and chemical composition of lettuces and tomatoes was investigated by Kalemabsa *et al.* (1998a). Vermicompost and ammonium nitrate were each applied at 4 rates. Vermicompost and FYM increased the yield of lettuces and tomatoes in comparison to the unammended control. Except at the lowest rate, ammonium nitrate decreased the yield of lettuces and tomatoes. In another experiment, Kalembasa *et al.* (1998b) studied the influence of same forms of N applied separately or in combination on yield and chemical composition of radishes and sweet peppers. They reported that the 1:1 mixture of vermicompost and ammonium nitrate increased sweet pepper yields but had no significant effect on radish yields. Kiepas *et al.* (1998) studied the effect of two vermicompost, produced from domestic wastes, on the growth of tomato cv. Remiz, plants in greenhouse experiments. The plants were grown in peat container media containing 0, 50 or 100 per cent of vermicompost. Both vermicompost contained a high level of available nutrients and salts. In container media, the concentrations of nutrients were raised with mineral fertilizers to obtain optimum plant growth. Seed germination and the growth of tomato plants in media amended with 50 per cent vermicompost were not significantly different to those in the peat control. Vermicompost delayed germination and decreased the fresh weight of plant in the early stages of growth, probably due to high salinity. However, there was an improvement in plant growth in media with vermicompost after 2 months. There were no symptoms of nutrient deficiency on these plants.

In a field experiment in 1997-98 in Assam, tuberiets (5-10 g) from true potato seed line HPS II/67 were grown at 50 x 10 or 20 cm spacing and supplied with the recommended NPK fertilizer rate (120:100:100 kg NPK/ha) or 50 or 75 per cent of the recommended NPK rate + 2.5 t vermicompost per hectare by Mrinal (1998). Row spacing of 50 x 10 cm produced higher marketable tuber yield than the wider spacing (12.0 Vs. 10.6 t/ha). The application of 75 per cent of the recommended NPK rate + 2.5 t vermicompost produced the highest marketable tuber yield of 14.1 t. Ramchandra *et al.* (1998) reported that the plant height at harvest, days to initiate flowering, number of branches per plant, number of pods, number of seeds per pod and yield of peas were highest with 10 t vermicompost + 100 per cent recommended NPK, amongst the various treatment combinations receiving 0, 50 and 100 per cent recommended dose of fertilizers and 0, 5 and 10 t vermicompost per hectare. Reddy *et al.* (1998) studied the effect of different levels of chemical fertilizers and vermicompost on yield and quality parameters of Coriander plants grown in alkaline soil. The control was provided with N, P205 and K20 at 100, 50 and 50 kg/ha, respectively. Vermicompost was applied at rates equivalent to 5, 10 or 15 t/ha. Plants were harvested 30, 35, 40 and 45 days after sowing for analysis of trace elements and ascorbic acid. Vermicompost at 10 or 15 t/ha significantly increased the concentrations of trace elements (Fe, Mn) in leaves. Ascorbic acid content was similar for control and in soil with vermicompost at 10 t/ha. Zn content was highest with 5 t vermicompost per hectare and declined with higher rates, but at all rates it was higher than in the control. Ascorbic acid content increased with delay in harvesting; trace elements generally increased upto 40 days and then declined.

Shashidhara *et al.* (1998) in a field experiment studied the effects of organic and inorganic fertilizers on growth and yield of chilli (Capsicum sp.) during Kharif for 2 years (1994-95). Treatments comprised organic sources like FYM (5 t/ha), vermicompost (2.5 t/ha), red gram stalk and biogas slurry (5 t/ha) in combination with 0, 50% or 100% of the recommended dose of fertilizers (RDF N: P205: K20 at 150: 75: 75 kg/ha). Organic sources had no significant influence on dry pod yield. Application of 100 per cent recommended dose of fertilizers together with organic sources increased dry pod yield (mean 639 kg/ha) significantly over 50 per cent and 0 per cent recommended dose of fertilizers (558 and 506 kg/ha, respectively). Srivastava (1998) reported that performance of $2/3^{rd}$ NPK supplied to maize through fertilizer + $1/3^{rd}$ N supplied through vermicompost was about equal to 100 per cent NPK supplied through fertilizer. Vadiraj *et al.* (1998) studied the response of three cultivars of coriander to graded levels of vermicompost in comparison with chemical fertilizers. Results indicated that application of vermicompost significantly increased herbage and seed yield and was comparable to applying chemical fertilizers. The herbage yield was greatest in cv. Rcr-41 (6067.5 kg/ha) at 60 days after sowing when 15 t/ha of vermicompost was applied. Seed yield was greatest in Rcr-41 (1314 kg/ha) from plants treated with 20 t/ha vermicompost. Yadav (1998) observed that the differences in grain yield of rice between the treatments receiving 100 per cent recommended dose of NPK and vermicompost @ 10 t/ha + 50 per cent recommended dose of NPK were not statistically significant, indicating possibility of substitution of 50 per cent of NPK dose. However, the significant increase in fruit yield of chilli to the extent of 5 q/ha was observed in the treatment receiving vermicompost @ 10 t/ha + 50 per cent recommended dose of NPK over the treatment receiving 100 per cent NPK dose only (48.1 q/ha) in rice - chilli cropping system.

Vasanthi and Kumarswamy (1999) used vermicompost prepared from organic materials @ 5 and 10 t/ha with N, P and K at recommended levels to rice crop. The results showed that the grain yields were significantly higher in the treatments that received vermicompost from any of the above organic materials plus N, P and K at recommended levels than in the treatment that received N, P, K alone. It was found that vermicompost @ 5 t/ha would be sufficient for rice crop when applied with recommended level of N, P and K. In field experiments in 1995-97 at Junagadh, sugarcane cv. COC-671 was given 25 t farmyard manure per hectare, 5 t green manure per hectare, 5 t vermicompost per hectare or 2 kg azotobacter per hectare in combination with 100 per cent or 50 per cent of the recommended inorganic NPK fertilizers. Highest cane yield of 43.87 t/ha was given by application of farm yard manure + 100 per cent recommended NPK rate which also gave the highest commercial cane sugar yield of 3.75 t/ha (Mathukia *et al.*, 1999). Dademal (2000) observed significant increase in fruit yield of okra by 14 per cent due to application of vermicompost @ 1.5 t/ha over the treatment receiving no manure and fertilizers.

Kadam (2000) studied the effect of vermicompost with and without inorganic fertilizers on yield, quality and mineral nutrition of cowpea - cowpea cropping sequence. He reported that integrated use of vermicompost @ 3.75 t/ha in conjunction with 25 per cent of the recommended dose of fertilizers (N @ 6.25 kg and P205 @ 12.5 kg/ha) was adequate and optimum to cowpea crop in respect of yield (Table 14). Application of vermicompost @ 10 t/ha alone and 100 per cent NP + vermicompost @ 5 t/ha significantly improved the protein and methionine content of grains of cowpea. Integrated use of recommended dose of fertilizers and vermicompost @ 5 t/ha resulted into maximum uptake of major nutrients such as N, P, K, Ca, Mg and S in comparison to application of either recommended dose of fertilizers or vermicompost (5 t/ha) alone (Table 15).

Dhopavkar (2001) compared the effect of various organic manures including vermicompost with chemical fertilizers on

Table 14: Effect of vermicompost and fertilizers on grain and stover yield (q/ha) of first and second crop of cowpea

Tr. No.	Treatments	Grain		Stover	
		First crop	*Second crop*	*First crop*	*Second crop*
T1	100% NP (N @ 25 + P205 @ 50 kg/ha)	11.81	12.32	10.50	10.80
T2	VC @ 5 t/ha	9.55	11.40	10.20	10.60
T3	100% NP + VC @ 5 t/ha	16.19	17.40	14.80	15.50
T4	75% NP + VC @ 1.25 t/ha	12.77	14.13	11.20	12.00
T5	50% NP + VC @ 2.50 t/ha	13.20	14.43	12.00	12.30
T6	25% NP + VC @ 3.75 t/ha	15.80	17.00	13.60	14.00
T7	VC @ 10 t/ha	14.87	16.53	13.00	13.40
	Mean	13.45	14.75	12.19	12.66
	S.E.±	0.66	0.48	0.35	0.69
	C.D. (P=0.05)	2.05	2.48	1.09	2.14

Table 15: Effect of vermicompost and fertilizers on total uptake (kg/ha) of nutrients by first and second crop of cowpea

Sr. No.	Treatments	*Nitrogen*		*Phosphorus*		*Potassium*		*Calcium*		*Magnesium*		*Sulphur*	
		First crop	*Second crop*	*First crop*	*Second crop*	*First crop*	*Second crop*	*First crop*	*Second crop*	*First crop*	*Second crop*	*First crop*	*Second crop*
T1	100% NP (N @ 25 kg+P205 @ 50 kg/ha.)	57.8	64.7	5.01	5.73	29.5	33.6	11.0	11.6	7.30	8.30	3.23	3.66
T2	VC @ 5 t/ha	49.2	57.7	4.36	5.73	25.7	32.6	11.5	11.5	6.80	8.30	3.11	3.75
T3	100% NP + VC @ 5 t/ha	84.6	96.2	9.96	11.09	48.6	58.9	16.4	19.4	12.4	14.3	4.82	6.34
T4	75% NP + VC @ 1.25 t/ha	62.2	72.2	5.90	6.84	35.9	42.1	11.0	13.1	8.70	10.1	3.45	4.26
T5	50% NP + VC @ 2.50 t/ha	64.5	72.9	6.29	6.88	34.3	42.8	10.9	13.5	10.0	10.7	3.47	4.55
T6	25% NP + VC @ 3.75 t/ha	74.5	84.5	8.43	9.39	37.9	47.7	12.2	15.7	12.1	13.2	4.72	5.66
T7	VC @ 10 t/ha	72.5	84.8	7.25	9.53	36.0	48.9	15.9	17.3	11.9	13.2	4.20	5.39
	Mean	66.7	76.3	6.74	7.87	35.4	43.9	12.7	14.6	9.9	11.1	3.86	4.80
	S.E.±	2.5	2.2	0.28	0.26	2.9	2.7	0.5	0.6	0.7	0.5	0.27	0.19
	C.D. (P=0.05)	7.8	6.9	0.87	0.79	8.80	8.2	1.5	1.9	2.3	1.5	0.83	0.60

VC – Vermicompost N – Nitrogen P – Phosphorus

Kadam (2001)

Table 16: Effect of organic manures and fertilizers on quality parameters of Alphonso mango

Sr. No.	Treatments	Ascorbic acid (mg/100g)	Reducing sugar (%)	Total sugar (%)	Total soluble solids (%)
1.	Vermicompost	72.083	3.079	11.180	14.760
2.	F.Y.M.	63.750	3.039	10.930	14.060
3.	Town compost	72.917	2.944	10.325	13.063
4.	Poultry manure	72.083	2.667	9.987	12.963
5.	Glyricidia	65.000	2.894	10.313	13.207
6.	Recommended dose	60.000	2.965	11.284	14.520
7.	Control	59.167	2.648	10.065	12.577
	S.E. ±	N.S.	0.047	0.201	0.453
	C.D. (0.05%)	–	0.146	0.619	1.395

Dose of manures was applied @ 0.75 kg N/tree

Recommended dose: 1.5 kg N + 0.5 kg P + 0.5 kg K + 50 mg FYM/tree

Dhopavkar (2001)

the yield and quality parameters of Alphonso mango in lateritic soil. He has reported significant improvement in quality parameters of mango pulp due to vermicompost application.

In situ application of vermicompost and crop productivity

Khasnitz (1922) observed that addition of live worm to a garden soil increased yield of peas and oats by 70 per cent. The body of worm contains upto 72 per cent of its dry weight as protein and it has been calculated that the body of a single dead worm can yield as much as 10 mg of nitrate nitrogen (Lawarence and Miller, 1945). Formation of aggregates is of prime importance, as soil rich in aggregates remains well aerated and drained by earthworms. In an experiment, the percentage of aggregates in soil to which earthworms were added as compared with that in soil without earthworms. After 3 days, the soil with earthworms and 12 per cent or more large aggregates, whereas, the soil without worms contained only 5 to 9 per cent aggregates (Hopp and Slater, 1949). It has been demonstrated by many workers (Rhee, 1965; Aldag and Graft, 1974; Altavinyte, 1974 and Altavinyte and Zimkuviene, 1985) that earthworms have beneficial effect on soil leading to increased yield of crops. Some of the effects of earthworms on soil take much time to show perceptible influence on plant growth. Large increased in yields of grass and clover were obtained when the soil was inoculated with earthworms as compared with control, using unproductive sub soil. A comparatively new angle is claimed that certain beneficial chemicals are released from the bodies of earthworms which increased crop yield (Hopp and Slater, 1949).

Earthworms improve the aeration of soil by their burrowing activity. They also influence the porosity of soil. Earthworm activity increased the porosity of two soils from 27.5 to 31.6 per cent and 58.8 to 61.8 per cent, four to ten times faster than the soil without earthworms (Nijhawan and Kanwar, 1952 b). Earthworms are observed to improve the nature of soil by breaking up organic matter and increasing the amount of nitrogen made available to plants either by excretion or decay of earthworm corpses (Barley and Jennings, 1959 and Satchell, 1967). Mackey *et al.* (1962) found that earthworms stimulate P uptake by crop due to redistribution of organic matter. The ratio of carbon to nitrogen in organic matter added to soil is very important. Plants can not assimilate mineral nitrogen unless this ratio is of the order of 20:1 or lower. The C:N ratio of freshly fallen litter is much higher than this critical parameter (Witamp, 1966). Earthworms increased the quantity of mineralized nitrogen and make it available for plant growth. Their corpses decay rapidly in a typical experiment, they had disappeared completely from soil after two or three weeks at 21 ^{0}C. Of the nitrogen added to the soil from the decomposed worm tissue, 25 per cent was in the form of nitrate nitrogen, 45 per cent unaccounted for probably consisted of undecomposed remain of setae and cuticle and microbial protein (Satchell, 1967).

The farmers following organic farming aim at *in situ* development of earthworms in gardens and orchards where the land is not ploughed often. The organic matter mulch is maintained at the base of the plants and drip/sprinkler irrigation is practiced by these farmers. Most of these farmers developed earthworms in bins and later as the population of earthworms started increasing, they released them into plant basins. The synthetic pesticides and chemical fertilizers are not used by these farmers. In such fields, these earthworms are thriving well and the soil living earthworms are also establishing with the formation of organic layer at the top. Coconut gardens, fruit orchards and some mulberry gardens are showing good response to this kind of practice (Kale, 1998). In a series of pot and box experiments, a strong correlation between the number of earthworms and the growth of barley has been demonstrated by Atlavinyte, Bydonavicience and Bada Vicience (1968) and Atlavinyte (1971). Aldag and Graff (1974) compared the growth of oat seedlings on brown podsol that had been treated with *Eisenia foetida* for 8 days with that in the same soil without worms. The dry matter yield of the oat seedlings was 7 - 8 per cent greater in the soil with earthworms and the total protein yield was 21 per cent more. Stockdill (1982) observed greater thatch incorporation after introducing *A.caliginosa* to pastures in New Zealand. The beneficial effect of the worms did not seem to be due to the cementation of casts by either calcium carbonate or intestinal mucus, for if it were so, the casts from both arable and grasslands should have been equally stable. Earthworm population was 3.4 times greater under no till inoculated plots compared with conventionally tilled inoculated plots (Sharma and Madan, 1983).

Earthworm channels are generally larger than root channels of annual plants and perennial grasses and worms are able to exert greater pressure than roots in forming channels (Mckenzie and Dexter, 1988a and 1988b). Lavelle (1988) discussed the food resources in the soil in relation to composition and earthworm utilization. Kale *et al.* (1989) observed that out of six species of earthworms only *Curgiona narayani* (Michaelson) could tolerate the waterlogged condition prevalent after transplanting of the crop. With regard to microbial population, worms had little effect on total bacterial population, but had marked influence on the number of N fixing bacteria and fungi. The density of the surviving worm population had no adverse effect on the productivity of soil. Earthworm channels sometimes promote rapid infiltration rate (Ethlers, 1975; Edwards and Lofty, 1980 and Zachmann *et al.*, 1987). Earthworm activity can potentially increase crop growth for many reasons e.g. increased nutrient uptake, improved soil physical properties, better mixing of the soil, increased infiltration rate (Datta, 1948). One of the reasons most often suggested is improved root growth. Roots tend to follow the pathways of least resistance, whether these are biopores, cracks, ped faces, or packing voids. Cylindrical biopores are often more stable than cracks and fine planes between peds. Stability of biopores increased with pore diameter and with more nearly vertical orientation of pores (Blackwell *et al.*, 1990). Basker *et al.* (1992) conducted an incubation experiment under controlled laboratory conditions which showed that exchangeable K was significantly higher in soil with earthworms than without earthworms. They concluded that the increase was due to the release of K from the non-exchangeable K pool as soil material was passed through the worm gut.

Edwards and Bater (1992) investigated the effects of inoculating earthworms of the species *Lumbricusterrestris* L.,

Aporrectodea longa (Ude), *Aporrectodea caliginosa* (Sav) and *Allolobophora chlorotica* (Sav) into intact soil profiles in the laboratory, plots on direct-drilled arable land in the field and newly capped waste disposal sites that had few or no earthworms. In all these studies, the earthworms increased significantly in number and the rate of growth and yield of plants grown on the inoculated sites were observed to be enhanced. Channels of the deep burrowing, surfaced feeding earthworms (i.e. *Lumbricus terrestris*) tend to be vertical. In contrast, most common species of earthworms from meandering temporary burrows at shallow depth, not permanent channels. Pot culture studies often indicated increased plant growth in response to unrealistically high earthworms inoculation rates. However, only a few field studies indicated any plant growth increases, mainly for pastures, orchards and small grains. Earthworms may improve crop growth without a direct influence on root growth. One benefit is the physical mixing of soil giving a more uniform distribution of residue and nutrients (Logsdon and Linden, 1992). In the experiment conducted by Hauser (1993), it was observed that earthworm surface casting activity was monitored in *Leucaena leucoccephala* alley cropping and a no-tree control system. Under *L. leucocephala* hedge rows, five times more casts (116.8 Mg/ha/year) were deposited than in the inter-row space (24.3 Mg/ha/year). *Hyperiodrillus africanus* and *Eudrilus eugeniae* were dominant species in alley and no-tree control plots, respectively. Amount of nutrients returned to the soil surface by casting were lowest in the no-tree control due to low nutrient concentration in casts of *Eudrilus* sp. In alley cropping (weighted average of amounts under hedge row and inter-row space), three times more N, K, Ca and Mg were recycled to the surface than in the no-tree control plots. The relative contribution of casts to nutrient cycling in alley cropping (nutrients in *L.leucocephala* pruning plus casts = 100%) was 33 per cent N, 16 per cent P, 6 per cent K, 16 per cent Ca and 34 per cent Mg. Permanent shading of the vicinity of hedgerows was identified as the most important factor enhancing casting activity. Basker *et al.* (1994) studied changes in potassium availability and other soil properties due to soil ingestion by earthworms and reported an increase in available non-exchangeable K in the cast of *L.rubellus* from Milson soil than for the non-ingested Milson soil. The exchangeable Ca was also higher as compared to the control soil.

Many organisms are responsible for decomposing organic matter that reaches the soil. Soil microorganisms are the major decomposing agents that breakdown plant and animal residues into useable nutrients for plants and other soil organisms. Not all plant materials are subjected to immediate decomposition by soil microorganism. Earthworms by their feeding activities and moving of plant and animal debris, contribute to this process. Earthworms derive their nutritional requirements by feeding on plant litter as well as a wide range of decaying organic substances. Earthworms play a key role in removing plant litter, dung and other organic material from the surface of the soil and communitive processes within the soil matrix. Typically, surface feeders are represented by *L. terrestris*. Earthworms in this group are primarily responsible for the transportation of organic material from the surface layer into the soil matrix. In the absence of these earthworms, surface mats may develop, locking up large quantities of nutrients. Earthworms that feed within the soil volume are represented by *A. trapezoids*, *A. tuberculata, A. turgida* and *O. tyaeum*. Earthworms in this group feed on the partially decomposed residue and microbes associated with the decomposing material. This suggests that residue decomposition may be controlled through comminution and microbial population regulation. Earthworms that group feed in the soil volume may be important in regulating microbial populations and thus, the rate of residue decomposition (Berry, 1994).

Agasimani *et al.* (1994) reported that application of vermicompost @ 1 t/ha and release of earthworms @ 50000/ha found better, respectively by giving 14 per cent (3785 kg/ha) and 13 per cent (3733 kg/ha) higher pod yield when compared with control (3316 kg/ha). Among the main plot treatments, application of FYM @ 4 t + 50 per cent recommended dose of fertilizes (RDF)/ha recorded significantly higher pod yield (4002 kg/ha) and higher number of earthworms (68/sq.mt.) compared with other treatment combinations. Palanisamy (1996) found an increase in yield of barley, oat and rape with increasing number of earthworms in the Temperate Zone. He also noted more absorption of nitrogen and phosphorus by these crops.

Kulkarni *et al.* (1996) showed in a field experiment the effect of vermicompost and in situ vermiculture on growth and yield of China aster cv. Ostrich plume mixed. The growth and flower yield at recommended dose of fertilizer + FYM, 75 per cent recommended dose of fertilizer + 2.5 t/ha vermicompost, 50 per cent recommended dose of fertilizer + 5 t/ha vermicompost and in situ vermiculture at 2 lakh earthworms per hectare were at par. By adopting vermiculture, he showed that vermiculture can reduce the usage of chemical fertilizer to the extent of 25 to 50 per cent and without application of any chemical fertilizer by adopting *in situ* vermiculture. An investigation was carried out in Raichur by Venkatesh *et al.* (1997) to study the influence of *in situ* vermiculture and vermicompost on yield and yield attributes of grape cv. Thompson seedless (2-year-old vines planted in 1992). The results revealed that application of 100 per cent of the recommended dose of fertilizers (RDF) produced the greatest yield which was significantly higher than that in all other treatments. However, in situ vermiculture and use of vermicompost with graded levels of chemical fertilizers, increased grape yields significantly over control (only mulched) and the sole application of vermicompost. The same trend was observed in case of flowering canes per vine, number of branches per vine and yield per vine. Jadhav *et al.* (1998) studied the influence of vermiculture in situ, vermicompost and NPK fertilizer on growth yield parameters of S-41 mulberry (*Morus alba*) in 1993-94. The recommended dose of fertilizers + FYM gave significantly higher growth parameters and leaf yield over rest of the treatments in first and second crops. However, from third crop onwards in situ vermiculture with earthworms @ 1 lakh/ha gave results at par as to that of recommended package whereas in fifth crop, earthworm at lower population (@ 50,000/ha) also resulted improved growth parameters and leaf yield. Effect of in situ vermiculture and vermicompost on availability and plant concentration of major nutrients in grape was studied by Venkatesh *et al.* (1998) in an experiment at Raichur in 1992, Thompson seedless grapes were treated with populations of 1000 or 2000 earthworms/ha to make vermicompost in situ or with 5 t/ha of vermicompost and 0, 25, 50 or 75 per cent of recommended NPK fertilizer rate of 300 kg N + 500 kg P + 1000 kg K/ha. All treatments increased available N, P and K in soil. The larger population of earthworms or vermicompost + 75 per cent of recommended NPK rate gave results comparable to application of the full recommended NPK rate.

Vermiwash its Methods of Preparation and its Chemical Composition

Vermiwash, a liquid collected after the passage of water through a column of worm action is very useful as a foliar spray to enhance the plant growth and yield and to check development of disease. It is a collection of excretory products and mucus secretions of earthworms, along with micro-nutrients from the soil organic molecules. Vermiwash if collected properly is a clear and transparent, pale yellow coloured fluid. The utility of vermiwash in aquatic productivity was found to be immense but its potential as a biocide, either singly or when mixed with botanical pesticides, was still under investigation. The vermiwash complex was observed to be efficient in raising of nurseries, lawns and orchids. (Ismail and Pramoth, 1995).

A system of vermiwash preparation described by Ramesh (1995) consisted of 30 cm. earthen pots filled with pieces of stones upto 10 cm height from the bottom, above which a filter paper or muslin cloth was spread. Then a layer of humus along with 150-200 worms was laid down. The hole situated at the bottom of pot was fixed with a rubber hose, through which vermiwash was collected. Everyday, two to three table spoons of fresh dung slurry was poured on the humus as feed for the worms. Also about 100 ml of water was added everyday from the top. As it passes through earthworm burrows, a part of it washes down and the gut of the worms as also castings. Everyday 100 ml of wash was obtained by this method. It was suggested to apply @ 100 ml/plant at the root zone or diluted 4 times and applied as foliar spray.

Ismail (1997) has suggested a method of preparation of vermiwash using a barrel in which the layers of cattle dung and hay were placed on the top of the layer of soil. The epigeic earthworms were introduced to produce compost at faster rate and anecics were put to produce a large number of drilospheres. The vermiwash was collected from the bottom of the barrel after sprinkling water through perforated mud or metal pot from the above. It was suggested to use the vermiwash for foliar spray either as such or dilution with water or 10% cow's urine. The physico-chemical characteristics of vermiwash given by him were as follows: pH-6.9, dissolved oxygen - 1.14 ppm, alkalihity - 70.00 ppm, chloride - 110.00 ppm, sulfates - 177.00 ppm, inorganic phosphate - 50.9 mg/l, ammonical nitrogen - 2.00 ppm, potassium - 69.00 ppm and sodium 122.00 ppm. Grappelli *et al.* (1985) noted growth regulators content in worm castings as follows: Gibberellins (GA3) - 2.75 mg/g, cytokinins (IPA) - 1.05 mg/g and auxins (IAA) - 3.80 mg/g.

One of the methods described by Kale (1998) consisted of an outer and an inner vessels. The inner vessel would have an outlet at the lower side of the vessel. The inner vessel was filled with decomposing organic matter and about 1 to 2 kg earthworms were accommodated in 12 to 16 litre capacity vessel. As the earthworms were accommodated in waste, water was slowly added into vessel in excess. The excess water flowing out through the outlet as thick syrupy fluid was collected in the outer vessel. The fluid so collected was siphoned out and after diluting, was used as foliar spray to different crops. In another method, the cement tank was built at an elevated place from which they want to collect the wash. The slope provided in the tank provided scope for excess water to flow out in drops as thick syrupy emulsion through a small outlet. This was collected in a container and stored in bottles. Before using, it was diluted and sprayed to crops.

Karuna *et al.* (1999) also prepared earthworm exudate and used successfully for foliar spray to Anthuriums. In this method, clitellated earthworms of Genus, *Edrilus eugeniae* were

A Vermiwash Unit

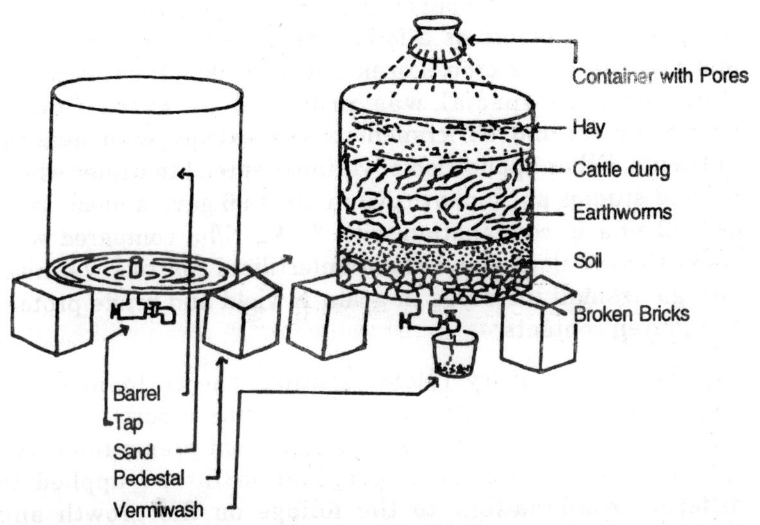

collected from the culture bins These worms were cultured on organic waste comprising mostly of leaf-litter, weeds, stubble mixed with one part of cow dung by volume in the form of slurry. Earthworms weighing one kg were placed into a dry enamel tray for 15 to 20 min to clear out the cast that will be excreted due to handling. Earthworms were then carefully separated from their excreta and then added into glass beaker having 500 ml of distilled water (37-40^{0}C luke warm). The worms were agitated for 3 to 5 min and removed and added into another bowl containing 500 ml of water at room temperature (25-27^{0}C) to rinse them thoroughly to collect the exudate adhering to its body wall before releasing back to the culture bins. The contents in the two bowls were mixed to use as a spray. The exudate thus collected was a syrupy, light yellow fluid.

Effect of Vermiwash on Yield and Quality of Crops

Kobatke (1954) reported that the coelomic fluid from earthworm body had antibacterial properties. Studies on effect of spraying of vermiwash on vegetables indicated that quality and quantity of yield were improved markedly. It was also observed that the foliage turned dense green in two to three days when spray was used on plants other than vegetables (Anonymous, 1993). Ismail (1997) reported that vermiwash can be sprayed on plants as a foliar spray for improving quality and yields of Okra crop. Lozek and Fecenko (1998) reported that vermisol special was produced by extraction of vermicompost and enrichment of the extract with mineral nutrients. Foliar application of vermisol special to winter wheat in field studies at Sladkovicovo in 1994-96 gave a mean yield of 7.62 t/ha in combination with 30 kg N/ha compared with 7.28 t from N alone and 6.71t in unfertilized controls. Vermisol also gave small increases in grain N yield and crude protein and gluten contents.

Snieg and Bury (1998) studied the influence of a biopreparation (a water extract of vermicompost), a water solution of urea, the vitamin-micro element preparation crop set and the micro element preparation Insol-7 applied in different combinations to the foliage on the growth and development of potatoes. The biopreparation generally had little effect on the formation of above ground plant parts, the assimilative leaf area and the content of chlorophyll and carotenoids. Insol-7 applied alone increased the mass of the above ground plant parts and crop set increased the leaf content of pigments. Karuna *et al.* (1999) worked on the two different concentrations of exudates of body fluid (vermiwash) of earthworm (*Eudrilus eugeniae*) and tested as spray to study its effect on long lasting inflorescence of anthuriums, which fetch a high premium in the international market. The results were compared with the treatments receiving 0.05% urea as spray during the course of experiment. It was found that the tested lower concentration of the vermiwash i.e. 50% was most effective in inducing vegetative growth like number of suckers, length and breadth of leaves and length of petiole. The vermiwash spray also initiated early flowering in plants. Lozek and Gracova (1999) reported that tomato seedlings (Cv. *Slava porynia*) planted in medium-heavy black soil in the green house were supplied with full N nutrition supplemented with 100 litres of vermisol (a vermicompost extract)/ha applied through the irrigation water 3 weeks after planting. Vermisol application increased yields by 7.3% and resulted in a decrease in fruit nitrate content by 15%.

Todkari (2001) studied the effect of vermiwash prepared by two methods viz. i) Percolating water through the drilospheres of the above mentioned decomposing organic residues from a vermiwash unit and ii) dipping and agitating one thousand earthworms in 1 litre luke warm distilled water for 5 minutes, on growth characteristics, yield and nutrition of three flowering plants. He inferred that vermiwash prepared by two different methods found to have good nutrient potential, as it contains 34, 13 and 90 ppm of average nitrogen, phosphorus and potassium respectively. (Table 17). For maximization of yield of flowers, the flowering crops like seasonal chrysanthemum and marigold be fertilized with recommended dose of NPK fertilizers and foliar spray of 100%. Vermiwash (II) and 100% vermiwash (I), respectively, while vermicompost @ 5t/ha along with recommended dose of NPK fertilizers be applied to china aster (Table 18 and 19). Foliar application of vermiwash prepared by both the methods resulted into considerable increase in total nitrogen phosphorus and potassium uptake by seasonal chrysanthemum, marigold and china aster in comparison to NPK alone which gave indications of quick absorption of the said nutrients through foliage for better nourishment of these flowering plants.

Table 17: Chemical composition of vermiwash prepared by two methods

Sr. No.	*Chemical properties*	*First method*	*Second method*
1.	pH	8.20	7.60
2.	N (ppm)	34	33
3.	P (ppm)	14	12
4.	K (ppm)	91	89

Balam (2000) studied the biopesticidal properties of vermiwash prepared from cow dung and vegetable wastes by inoculating *Eisenia foetida* species of earthworms in the laboratory and the effect of said vermiwash on the powdery

Table 18: Total dry matter yield (t/ha)

Tr. No.	*Treatments*	*Seasonal chrysanthemum*	*Marigold*	*China aster*
T1	Control	3.96	3.21	7.21
T2	NPK (@ 90:60:60 kg/ha for chrysanthemum and marigold) (150:60:60 kg/ha for china aster)	4.73	4.12	9.52
T3	NPK + Vermicompost @ 5t/ha	5.47	5.33	11.51
T4	NPK + 100% Vermiwash (I)	5.94	5.44	10.82
T5	NPK + 50% Vermiwash (I)	5.69	5.11	10.53
T6	NPK + 100% Vermiwash (II)	5.35	5.32	11.12
T7	NPK + 50% Vermiwash (II)	5.33	4.74	10.51
T8	NPK + 0.05% Urea solution	4.97	4.72	9.83
	C.D. (P = 0.05)	0.54	0.82	0.77

Table 19: Flower yield (t/ha)

Tr. No.	*Treatments*	*Seasonal chrysanthemum*	*Marigold*	*China aster*
T1	Control	1.56	6.33	2.11
T2	NPK (@ 90:60:60 kg/ha for chrysanthemum and marigold) (@ 150:60:60 kg/ha for china aster)	3.02	11.02	4.92
T3	NPK + Vermicompost @ 5t/ha	3.52	14.81	10.01
T4	NPK + 100% Vermiwash (I)	4.51	16.13	6.82
T5	NPK + 50% Vermiwash (I)	3.75	15.02	6.81
T6	NPK + 100% Vermiwash (II)	4.61	15.91	7.72
T7	NPK + 50% Vermiwash (II)	3.81	13.13	6.91
T8	NPK + 0.05% Urea solution	3.02	12.72	4.72
	C.D. (P = 0.05)	1.4	1.92	2.03

Todkari (2001)

mildew disease of cowpea. Vermiwash at 20 percent and 30 percent inhibited the mycelial growth of pathogenic test fungi. But is inhibition percentage was different for different pathogen. Vermiwash at 30% concentration gave more than 37 per cent inhibition of *Curularia pallescens* and 28 per cent inhibition of *Alternaria alternata.* In control of powdery mildew disease of Cowpea in field with vermiwash and vermicompost the treatment T6 (vermicompost + vermiwash) was found to be most effective giving 75.14% disease control. The next effective treatment was vermicompost + vermiwash + 10% cow urine which showed 73.37% disease control. The treatments receiving vermiwash and vermiwash + 10% cow urine resulted in 62.71% disease control followed by the treatment receiving potash @ 30 kg/ha and vermicompost with 40.83 and 39.64 per cent disease control, respectively.

Use of earthworms in land improvement and reclamation

It is now well established that the introduction of earthworms into soils that have no earthworms or have only low natural earthworm populations, is usually beneficial in terms of plant growth and crop yield. Stockdill (1959) and Stockdill and Cossens (1966) have successfully introduced earthworms into pastures in New Zealand that lacked native earthworms. Following establishment of the earthworm populations, the organic matter at the base of the grass disappeared and compacted soils with poor structure were transformed into deep friable top soils.

Some of the methods of earthworms inoculation for the improvement and reclamation of land tried by various workers are described below:

1. Direct Release of Worms

A large number of deep working species of earthworms are introduced by sufficiently moistening and liming of acidic soils. If food is a limiting factor, addition of organic material such as municipal sludge or animal wastes allow their earlier establishment to facilitate stabilisation of temperature and moisture conditions. The mature worms should be placed at the rate of $150/m^2$.

Addition of earthworms to soils seems particularly promising in reclaiming flooded areas that are subsequently drained and put into cultivation with tree crops. In such cases, earthworms can be introduced at the rate of 100-800 worms per tree in the basin around the base of tree and mulched with cattle dung and leaf litter.

2. *As Soil Blocks*

Earthworms can be transferred to new localities by moving blocks of soil cut from the surface at localities where target species are common. Such blocks of soil are commonly introduced into small areas and supplied with organic matter as food.

3. *As Turf Pieces*

In this method, turf pieces with earthworms are cut and placed on the ground surface at 10 m spacing between them. To cut turf pieces, sod machines are developed in New Zealand. This method is useful for introduction of earthworms into pasture devoid of earthworm before.

4. *As Vermicasting*

Surface application of worm casts to degraded soil can improve its structure and fertility. Such lands may be planted with grass cover for the establishment of earthworms. Preliminary work in our country revealed that saline lands can be reclaimed with planting selected species of trees and fertilizing them with vermicastings at the rate of 5 tonnes/ha. Over this layer, a 25 mm thick layer of cow dung was applied followed by a layer of mulch and oil cakes.

Barley and Keleinig (1964) successfully introduced *Aporrectodea coliginosa* and the megascolecid species, *Microscolex dubius* (Fletcher) into newly sown, irrigated pasture on sandy loam soil in Austrialia with corresponding significant improvements is soil structure, loss of organic materials and increased productivity. Ghilarov and Mamajev (1966) inoculated earthworms into reclaimed irrigated land in Uzbekistan and reported considerable improvements in soil structure and fertility. Van Rhee (1969 and 1971) introduced earthworms into polders that had been drained and reclaimed from the sea in the Netherlands and reported that this accelerated the development of normal soil structure and facilitate the returns of the polders to productive use. He found that grass yields of these polders increased upto four times, and clover yields upto ten times, after inoculation with earthworms. The dry matter yield of spring wheat was doubled by inoculation with earthworms and fruit trees were established much more readily than in uninoculated soil.

Edwards and Bater (1992) investigated the effects of inoculating earthworms of the species, *Lumbricus terestris* L., *Iaporrectodea caliginosa* (Sav.) and *Allolobophora chlorotica* (Sav.) into intact soil profiles in the laboratory, plots on direct-drilled, arable land in the field and newly capped waste disposal sites that had few or no earthworms. In all these studies, the earthworms increased significantly in number and rate of growth and yield of plants growing on the inoculated sites.

Earthworms for Animal Feed

The earthworm is a very promising source of high quality animal protein the chemical composition of earthworm meal is presented in Table 20.

Table 20: Chemical composition of worm meal

Sr. No.	*Chemical composition*	*Constitution in per cent*
1.	Moisture	7.70
2.	Crude protein	56.27
3.	Total ash	10.49
4.	Crude fibre	0.56
5.	Calcium	1.01
6.	Phosphorus	1.25
7.	Fat	6.05
8.	Silica	3.76

The protein content of worm meal is considerably greater than that of the fish meal and soybean meal proteins (Kale and Bano, 1984). Moreover, earthworm protein is an easily digestible protein as seen from the amino acid content of the worm meal (Table 21).

A lot of research is in progress in this country to serve worm meal as a protein rich supplement to poultry, aquaculture and domestic animals. Earthworms fed to chicks, crabs, fishes and tad poles have shown encouraging results for stimulating enhanced growth rate. It was noticed that when fish meal of the poultry feed was substituted with worm meal, there was no difference in the body weight of the broilers (Kale, 1985). In aquaculture, the type of food bears the greater potential effect on growth efficiency of fish. The food utilisation, growth and conversion efficiency were observed to be higher in commercially important catfish *Mostu vittatus* directly fed on earthworms (Arunachalam and Palanichamy, 1984). Srivastava (1984) reported that even gonodectomised and pinealectomised fish fed on the earthworms were quite healthy and thrived well. In U.S.A. and U.K. earthworms growth in sterilised filter paper are being utilized even as human food (Senapati and Pani, 1984).

Table 21: Amino acid content of worm meal of *Eudrilus eugeniae*

Amino acid	*In g/100 g protein*
Aspartic acid	10.3
Threonine	4.8
Serine	4.8
Glutamic acid	13.8
Alanine	5.2
Cystine	1.6
Valine	5.0
Methionine	2.0
Isoleucine	4.5
Leucine	7.9
Tyrosine	3.4
Phenylalanine	4.1
Lysine	7.1
Hystidein	2.6
Agrinine	6.1

Nutritional Status of Earthworm Protein

The earthworms have 72% crude protein which is much more higher than any other systems, worm meal is much better than yeast meal and fish meal. It has biological value of 84% and digestibility of protein 85% with net protein utilization of 79% and protein efficiency ratio of 4%. The amino acid composition is far superior to snail, meat and fish meal. For example Arginine is 4.13% in worm meal against 3.4% in fish meal. Likewise Tryptophan is 2.29% in worm meal as against 1.07% in meat meal and 0.80% in fish meal (Ghatnekar *et al.*, 1998).

Presumptive evidence from the successful poultry and pig feeding trials done by number of earlier research workers followed by numerous food chain studies, has proved that earthworms are unlikely to contain any naturally occurring toxins. A double blined feed trials on poultry and Chinese hamster proved that the worms procured from cellulose based industries waste are absolutely free of any other harmful contaminates and there was absolutely no side effect on the growth.

In fine, worm proteins have been universally accepted as concentrated feed for poultry, cattle and pets. It has tremendous market in aquaculture systems of Shrimps, Carps, Catfishes and Tilapias. Not only that worm protein has very good potential for human consumption. Depite some squeamishness, however several contests for recipes – using earthworm, have been held in Canada and US (Ghatnekar *et al.*, 1998).

Medicinal Value of Worm Meal

The harvested adult worms are to be washed thoroughly and left in clean containers on wet filter paper for 48-72 hours. The containers are to be covered tightly to prevent the worms from crawling out. The worm fed on filter papers and its gut gets cleared of the dung matter. Fresh filter paper is to be spread in the container every 24 hours. Once the worms start excreting only filter paper, they have to be removed and blotted gently with several layers of cloth, filter paper or towel. Then the worms are to be spread out on plastic sheet in shade to lose about 90 per cent of body water. Later, they are to be dried for 8 to 10 hours in hot air and ground to powder. The worm powder serves as 100 per cent worm meal. Worm tissue has tremendous medicinal value. This is known from ancient times and is being tested in recent times. Earthworm is certified for protein deficiency and utility in treatment small pox, teeth diseases, jaundice, rheumatism, for smooth delivery, gall bladder stone, during growth of hair and for arousing strong sexual desire, curing importance and scorpion and snake bite. Worm tissue is also antipyretic to bring down body temperature. In Japan, four kinds of drugs are known to be prepared from earthworms including antibioses, aphrodisiacs, antipyretics and antidotes for poisons. Earthworm is being utilized in medicinal research to study generation mechanism, cancer and birth control at some universities like California of U.S.A. (Misra, 1984).

Medical Application of Vermiculture Biotechnology

Folk history abounds with stories of the medicinal value of earthworms, dating from at least as back as 2600 B.C. and covering a range of diseases from pyorrhea to post partial weakness, from jaundice to increase in the sperm count and as excellent aphrodisiacs. In fact, even today the Japanese way they put them in hot sake (wine made from rice), allowing them to secrete a milky liquid and then drink the potent bur. They consider it as an aphrodiasic additive to sake, their white wine. Nevertheless, scientific experimentation inclusive of double blind clinical trials are quite few.

Vermiculture – waste Treatment Plants – salient Features:

Biotechnology Resource Centre (BRC) in last 14 years has dealt with effluents of several industrial such as: 1) Paper mills, 2) Soybean oil extraction plants, 3) Naptha based organo-chemical, 4) Dairy and cheese, lactose production from whey, 5) Tannery, 6) Sewage, 7) Pesticides, 8) Hospital and Surgical, 9) Enzyme production unit.

All these diverse industries treat their solid effluents by the vermiculture technology developed by BRC wherein their daily effluent generation is of the order of 5-10 tonnes. The resultant product after vermiculture treatment plant implementation is value added biofertilizer and protein biomass.

Apparently there is no criterion as to which wastes will be treated by earthworms. However, it can be generalised that the pH of the effluents should be within the range of 5.5-8.5 and the solid contents should be around 40-50%. Where the solid content is less it can be compensated many times by in house wastes inclusive of boiler or fly ash, lime sludge, other wastes such as paper clippings, canteen, kitchen and if possible even toilet wastes. Other cheap ingredients that could be mixed to increase solid contents could be any of the cellulosic wastes like old newspapers, sawdust, fresh or dry cattle dung, green foliage etc.

Vermiculture waste treatment technology is not very useful to convert wastes containing pieces of plastics, polythene and polystyrine foam, glass, iron or steel scrap. In fact, even small amounts of iron filings or broken glass pieces present in any type of wastes need to be manually separated out. The treatment plant consists of series of smooth walled cemented tanks with moisture regulatory mechanism and meshed covers.

Normally earthworms need to be acclimatized to particular wastes in stages and at times it may take about 4-8 months even. Afterwards however, the retention time or cycle period is reduced to 40-45 days. Inoculation of specific microorganisms like cellulose degraders is very helpful. Cellulose degraders can be of three main types such as fungal, actinomycal and bacterial. To give an example species of *Cellulomonas* and *Aspergillus* work very nicely with earthworms to degrade paper mill effluents. So paper mill effluents should be seeded with these cellulose degraders to stimulate the biodegradation by earthworms. Presence of cellulose degraders than along with worms bring about total degradation of effluent sludge by series of chain reactions such as prolific increase in nitrogen fixers, phosphate solubilizers and potassium enrichers. Thus the resultant biodegraded product becomes biofertilizer of excellent quality. Thus waste is indeed converted into wealth.

Most important thing that will be deciding factor for many industries to undertake vermiculture biotechnology is the space requirement. It could be generalized that to treat 5-7 tonnes of wastes every day specific industry may need an open space of 1000-1400 sq. mt. In this open space series of trenches or cemented, plastered tanks could be made depending upon the convenience and the type of wastes.

Apart from space another very important factor to employ vermiculture treatment is the costs. Though it is difficult to give exact estimate of the cost it can be generalized that against conventional biodegradation or chemical treatment plant that initial capital costs will be 1/5 or 1/6th only. Another startling advantage will be that running cost of vermiculture effluent treatment plant shall be almost negligible in terms of power consumption. On the other hand, value added biofertilizer is produced which can be a saleable product from Rs. 5/kg to Rs. 20/kg. Liquid formulation to mix with drip irrigation system or to be sprayed as foliar spray can be also made from cultured earthworm biomass which is also saleable from Rs. 60 to Rs. 92/litre. Cost of production and packing of solid biofertilizer can not come over Rs. 2/kg whereas that of liquid formulation not more than Rs. 20/liter. Many paper mills where this technology is transferred by BRC are selling 5-6 tonnes of biofertilizers daily.

Earthworm and Pollution Control

The urban and rural wastes being generated continuously are undesirable pollutants for the environment and a menace to the health of the community. Utilisation of earthworms for the disposal of such wastes is less cumbersome and costly. The sources of toxic substances reaching the soil surface are mainly the solid wastes containing heavy metals released from the industries and pesticides used for health and agriculture. In recent years, increasing amount of metals like mercury, lead, arsenic, cadmium etc. are discharged to the environment by industrial establishments. These metals progressively reach the human beings through food chain. The special characteristics of heavy metal chemicals are their strong attraction to biological tissues and the slow elimination from biological systems. Residues of heavy metals like cadmium, lead, zinc and nickel have been recorded in their bodies. The accumulation of toxic chemicals in earthworm tissues is very significant ecologically, because these animals are important components in the food chain of several species of birds and mammals. The earthworms serve as a part of diet of some birds like wood cock, robins and of reptiles (Misra, 1984). This is very important from the point of minimisation of soil pollution.

Earthworms can process household garbage, city refuse, sewage sludge and wastes from paper, wool and food industries. They dominate the soil invertebrate biomass in many ecosystems of the world (Atlavinyte and Pociene, 1973; Appelhol, 1980; Gaddie, 1980 and Mba, 1978). Dash (1978) suggested that earthworms can be utilized for decomposing waste organic biomass. Ireland (1975) observed that earthworm *D. rubida* living in heavy metal contaminated acid soil can increase he amount of available Zn and Cu by excretion in the faeces with little or no effect on the amount of available lead. Fileldson (1985) experienced that water pollution and odour problems can be controlled economically by earthworm culture.

The effects of chemical on earthworms and methods of investigating these effects have been reviewed by Satchell (1955), Davey (1963), Edwards and Thompson (1973) and Edwards and Bohlen (1992). Most reports in the scientific literature refer to the results of field testing of the effects of chemicals on earthworm populations. It is extremely difficult from the reports to assess reliable, the relative toxicity of different pesticides, due to the considerable variability between test sites, soils, size of plots, replications, formulations, doses, methods of application and crop grown. Although field tests have been adequate to identify those chemicals which are extremely toxic to earthworms, it is much more difficult to identify accurately, compounds which are only moderately or slightly toxic to earthworms.

The effects of pesticides on earthworms have been assessed in the laboratory by treating soils with chemicals and then adding test earthworms. Methods involving in voluntary ingestion of contaminated food material (Stringer and Weight, 1973 and Fayolle, 1979) provide results that are difficult in interpret and food repellency affects the valid assessment of mortality. Injection of toxicants into the body tissues of worms (Stenersen *et al.*, 1973 and Gilman and Vardanks, 1974) can cause direct injury and toxicant placement can be uncertain. The earthworm toxicity testing techniques were investigated experimentally in the laboratory at Rothamsted Agril. Research Station. These methods included: toxicity tests in natural soils (Hopkins and Kirk, 1957; Edwards and Lofty, 1972 and Ruppel and Laughlin, 1977), exposure in a silica gel paste-glass beall matrix (Bouche, 1988), immersion in chemical solution (Martin and Wiggans, 1959; Ghabbour and Iman, 1967 and Lehrun *et al.*, 1981) are forced feeding tests (Stringer and Wright, 1973).

Earthworms were selected as one of the five key indicator organisms for eco-toxicological testing of the toxicity of industrial chemicals, not only by the European Economic Community (E.E.C.) but also by the organization of the United Nations (F.A.O.) and by many national pesticide registration authorities and environmental pollution committees. The two standardised test methods developed at Rothamsted Experimental Station (Edwards, 1983 and 1984) are now used by these organizations. These include one exposing earthworms to chemicals in filter paper and one to chemicals in artifical soils are described and the median lethal concentration (LC50) for chloracetamide, pentachlorophenol, chlordane, carbaryl, potassium bromide, copper sulfate and trichloracetic acid, based on assays done in 34 laboratories.

Waste Management and Resource Recovery

Agricultural, food processing or any cellulose based industry generates massive quantities of solid and liquid wastes. These wastes or as they might be more appropriately called 'Misplaced Resources', present a bi-fold problem. On the one hand they are unbelievably noxious, or likely to become so, and problems inherent in their disposal can represent considerable financial burden to the industry generating them. On the other hand they are a rich cellulose based resource that is currently wasted-large quantities of nutrients, energy proteins, carbohydrates, most often in forms that have taken considerable time, effort and money to produce are now being abandoned.

Vermiculture systems developed by the *Biotechnology Resource Centre* is able to contribute to the solution of both these problems.

Vermiculture aids in disposal by significantly improving the physical qualities of waste and reducing smell and putrescent potential. Such culture also recaptures much of viable products *viz.*, 1) A residue vermicompost that is of considerable value in the horticulture, floriculture, silviculture systems as a potting soil, organic biofertilizer, soil conditioner and plant tonics and 2) the worms themselves which because of their very high protein content have considerable potential as protein additive to the ration of both domestic and exotic livestock.

At all stages in food production – on the farm, processing, distribution and consumption – considerable biological losses occur. For instance, of the dietary protein eaten by a meat chicken wholly some 12.5% is totally purchased as a broiler chicken by the consumer and more of this is lost in preparation, cooking and consumption.

Likewise losses in production and particularly processing of non-food agricultural items, e.g. wool, leather can also be extensive. Given the earthworms itself (on dry wt. Basis) is approximately 72% protein, then the question arises as to whether or not earthworms might be used to harvest these protein losses. Several attempts have been made to calculate the production potential for earthworm protein.

The calculated figures are very high, greatly in excess of any other form of animal protein production, the important question now is can such a production system work economically in practice? The variability of vermiculture as an option for waste management dissolves around 3 possible financial implications, of which protein production might be only the third. The disposal of most agricultural and cellulosic wastes represents substantial cost of the producer and it might be achieved by the vermiculture operation at no cost or even negative cost. Nevertheless, our experience based on our full scale trials in cellulose based industries suggests evidently that the premium financial returns is likely to come form the sale of vermicompost – a high grade nutrient rich compost of significant value as an organic biofertilizer, soil conditioner and tonics for plantations. It must be remembered however the quality of vermicompost depends upon the nature of starting waste, species of earthworm and overall efficiency of vermiculture system.

As such biofertilizers made from vermiculture have tremendous applications in all the key developing countries. With tremendous crisis of chemical fertilizers in most of developing countries including India, biofertilizers and plant tonics from vermiculture is indeed key to overcome energy crisis, maintain the environment and improve the economy. Raw materials of vermiculture like biodegradable celluloses coming from various paper mills, saw mills, dairies, soybean oil extraction plants is plentily available in our country. In fact, most of these paper mills and wood mills have problem of disposing solid effluents in the form of semi-solid sludge even after considerable amount of such effluent is recycled. Every day even small paper mills (30 t/day capacity) produced upto 0.75-1.25 tonnes of solid effluent which is normally dumped.

REFERENCES

Agasimani, C.A.; Patil, R.K.; Ravishankar, G. and Mannikeri, I.M. (1994). Effect of manures and fertilizers on activity of earthworms and productivity of kharif groundnut. *Groundnut News*, 6(2): 5-6.

Albanell, E.; Plaixats and Cabrero, T. (1988). Chemical changes during vermicomposting (*Eisenia foetida*) of sheep manure mixed with cotton industrial wastes. *Biology and Fertility*. 6(3): 266-269.

Aldag, R. and Graft, O. (1974). Z. Landw. Forsch. 31(11): 277-284. In: Sharma, N. and Madan, M. (1983). Earthworms for soil health and pollution control. *J. Sci. Ind. Res.* 42: 575-583.

Altavinyte, O. (1974). *Inst. Zool. Parasit. Acad. Sci., Lithuania*. 1(65): 69-79.

Altavinyte, O.; Baydonaviciea, Z. and Band – Vicience, I. (1968). Pedobiologia, 415-423. In Sharma, N. and Madan, M. (1983). Earthworms for soil health and pollution control. *J. Sci. Ind. Res.* 42: 575-583.

Anonymous (1992). Recycling of crop residues in the soil and its effect on performances of rabi sorghum. 29th meeting of research review committee. *Report Agril. Chem. and Soil Sci.* 1992-1993, M.P.K.V., Rahuri, India.

Anonymous (1993). Vermiwash promotes crop growth. *Indian Coconut Journal.* 1: 4.

Anonymous (1996). A Report of the Sub-Committee meeting of Agronomy and Agril.Chemistry and Soil Science of Konkan Krishi Vidyapeeth, Dapoli.

Anonymous (1998b). A newsletter of *INORA* 2(2): 3.

Appelohof, M. (1980). *Celloquinum of Soil Zoology, EPA, USA*: 157-160.

Arunachalam, S. and Palanichamy, S. (1984). Earthworms as a feed for the catifish *Mystus vittatus. A paper presented to National Seminar on organic waste utilization and vermicomposting, School of Life Sciences, Sambalpur University Jyoti Vihar, Orissa.* p. 4.

Atlavinyte, O. (1971). Pedobiologia, 104-15 In: Earthworms for soil health and pollution control. Sharma N. and Madan, M. (1983). *J.Sci.Ind.Res.* 42: 575-583.

Atlavinyte, O. and Zimkuviene, A. (1985). Effect of earthworm on barley crop in the soils of various density. *Pedobiology*. 28: 305-310.

Atlavinyte, O.; Daciulyte, J. and Lugausks, A. (1971). Correlations between the number of earthworms, micro-organisms and vitamin B12 in soil fertilized with straw. *Liet TSR MoksluAzad Darb Ser.* B:3:43-56.

Atlkavinyte, O. and Pociene (1973). *Pedbiologia*. 13: 445-455.

Balam (2000). Studies on biopesticidal activity of vermiwash in control of some foliar pathogens. M.Sc. (Agri.) Thesis submitted to Dr. B.S.K.K.V., Dapoli.

Bano, K and Kale, R.D. (1984). *A paper presented to National Seminar on organic waste utilization and vermicomposting. School of Life Sciences, Sambalpur University, Jyoti Vihar, Orissa*: 5-8 December.

Bano, K. and Kale, R.D. (1992). Potentials of earthworm farming Proc. *National Seminar on Organic Farming, M.P.K.V., College of Agriculture, Pune*, pp. 26.

Barley, K.P. and Jennings, A.C. (1959). Earthworm and soil fertility III. The influence of earthworms on the availability of nitrogen. *Aust. J. Agr. Res.* 10: 364-370.

Barley, K.P. and Kleining, C.R. (1964). The occupation of newly irrigated land by earthworms. *Australian Journal of Science.* 26: 290.

Basker, A.; Kirman, J.H. and Macgregor, A.N. (1994). Changes in potassium availability and other soil properties due to soil ingestion by earthworms. *Biol.Fertil.Soils.* 17: 154-158.

Basker, A.; Macgregor, A.N. and Krirman, J.H. (1992). Influence of soil ingestion by earthworms on the availability of potassium in soil. An incubation experiment. *Biol.Fertil.Soils* 14: 300-303.

Bawa, J.N. (1995). Studies on the effect of different organic manures, chemical fertilizers and growth promoter on growth, yield and quality of rabi-hot weather groundnut (*Arachis hypogaea* L.) under lateritic soil of Konkan region. M.Sc.(Agri.) Thesis submitted to Konkan Krishi Vidyapeeth, Dapoli. (Unpublished).

Berry, E.C. (1994). Earthworm and other fauna in the soil. *Advances in Agronomy*, pp. 76-77.

Bhangrath, P.P. (1996). Vermicomposting of organic residues by pit method. M.Sc. (Agri.) Thesis submitted to K.K.V., Dapoli.

Bhawalkar, V.U. (1992). Recycling of sugar residues by vermiculture biotechnology. *Proceeding of the National Seminar on Organic Farming held on April, 18-19, 1992*, pp. 53-54.

Bhawlkar, U. and Bhawalkar, V. (1992). Vermiculture biotechnology Bhawalkar *Earthworm Research Institute, A/3, Kalyani, Pune – Satara Road, Pune,* 411 037.

Bhidey, M.R. (1994). Earthworms in agriculture. *Indian Farming.* 43(12): 31-33.

Bhole, R.J. (1992). Vermiculture Biotechnology Basis, scope for Application and Development. *Proc. National Seminar Organic Farming held at College of Agriculture, Pune, January, 1992*: 28-29.

Blackwell, P.S.; Green, T.W. and Mason, W.K. (1990). Responses of biopore channels from roots to compressio by vertical stresses. *Soil Sci.Soc.Am.J.* 54: 1088-1091.

Bouche, M.B. (1977). In soil organisms as components of ecosystems (Ed. Lohm, U., and Persson, T.), *Econ. Bull.* 25: 122-132.

Bouche, M.B. (1988). Earthworm toxicological tests, danger assessment and biomonitoring a methodological approach. *In Earthworm in*

Waste and Environmental Management. (CA Edwards and E.F. Neuhauser Eds.) pp. 315-320, S.P.B. Academic, The Hague.

Chan, P.L.S. and Griffiths, D.A. (1988). Vermicomposting of pretreated pig manure. *Biological Wastes*. 24(1): 57-69.

Cheshire, M.V. and Griffiths, B.S. (1989). The influence of earthworm and cranefly larvae on the decomposition of uniformly 14C leveled plant material in soil. *J. of Soil Sci*. 40(1): 117-123.

Cooke, A. and Luxton, M. (1980). Effect of microbes on food selection by *L. terrestris Rev. Ecol. Biol. Sol*. 17: 365-370.

Dademal, A.A. (2000). Studies on nutrient management for okra with relation to its growth, yield response, nutrient uptake and changes in nutrient availability on lateritic soil of Konkan. M.Sc.(Agri.) Thesis submitted to Konkan Krishi Vidyapeeth, Dapoli (Unpublished).

Darwin, C. (1881). The formation of vegetable mould through the action of worms, with observation of their habits (Murray, London): 326.

Dash, M.C. (1978). Role of earthworm in the decomposer system. In Glimpses of ecology edited by J.S. Singha and B. Gopal (International Scientific Publication). 399-406.

Dash, M.C.; Mishra, P.C. and Behera, N. (1979). Fungal feeding by tropical earthworm. *Tropical Ecology*. 20: 9-12.

Datta, A.K. (1948). Earthworm and soil aggregation. *Amer.Soc.J.* 40: 410-437.

Davey, S.P. (1963). Effect of chemical on earthworms. A review of literature. Special scientific report, Wild Life No. 74. Washington.

Day, G.M. (1980). The influence of earthworms on soil microorganisms. *Soil Sci*. 69: 175-184.

Deshpande, M.S. (1992). Natural Agriculture and Earthworm. Abstracts of *National Seminar on Organic Farming. Abstracts published, M.P.K.V., College of Agriculture*, Pune, pp. 29.

Devi Dayal and Agarwal, S.K. (1998). Performance of sunflower hybrid as influenced by organic manures and fertilizers. *J. Oilseed Res*. 15(2): 272-279.

Devi Dayal; Agarwal, S.K. and Dayal, D. (1998). Response of sunflower to organic manures and fertilizers. *Indian J. Agronomy*. 43(3): 469-473.

Dharmalingam, C. (1996). Vermicompost pelleting enhances productivity in soyabean. In: Notes on National Training on vermiculture prepared by Directorate of Extension Education, Tamil Nadu Agril.University, Coimbatore, pp. 50.

Dhopavkar, R.V. (2001). Effect of organic manures on yield, quality and nutrient contents of mango (*Mangifera indica*) and forms of N, P and K in soil. M.Sc. (Agri.) Thesis submitted to Dr. B.S.K.K.V., Dapoli.

Dodalto, G.G. and Costa, L.M.D.A. (1990). Relationship between chemical characteristics of soil and worm cast (*Glossoscolex* spp.). *Rerista Gres.* 37(212): 331-336.

Dubash, P.J. and Ganti, S.S. (1964). Earthworms and amino acids in soil. *Current Science,* 33: 219-220.

Edwards, C.A. (1981). The use of earthworms in waste disposal and protein production. Rothamsted Expt. Stn., England.

Edwards, C.A. (1982). *Report Rothamsted Expt. Stn.* pp. 103-106.

Edwards, C.A. (1983). Earthworms, organic waste and food span. *Shell Chem. Co.* 26(3): 106-108.

Edwards, C.A. (1984). Report of the second stage in development of a standardized laboratory method for assessing the toxicity of chemical substances to earthworms, pp. 99, Report of Commission of European Communities, EUR 9360 EN.

Edwards, C.A. and Bater, J.E. (1992). The use of earthworms in environmental management. *Department of Entomology, the Ohio State University*, 1683-1689.

Edward, C.A. and Bohlen, P. (1992). The effect of toxic chemicals upon earthworms. *Review of Environ. Contam and Toxic.* 125: 23-99.

Edwards, C.A. and Lofty, J.R. (1972). *Biology of Earthworms*, Chapman and Hall, London.

Edwards, C.A. and Lofty, J.R. (1980). Effects of earthworms inoculation upon the root growth of direct drilled cereals. *J.Appl.Ecol.* 17: 533-543.

Edward, C.A. and Thompson, A.R. (1973). Pesticide and the soil fauna. *Residue Review*. 45: 79.

Edwards, C.A.; Burrows, I.; Fletcher, K.E. and Jones, B.A. (1985). The use of earthworms of composting farm wastes. *In composting of Agric. and other Wastes. Ed. J.K.R. Gasser. Elsexier Publ. Co., Amsterdam*, pp. 229-242.

Ehlers, W. (1975). Observation on earthworm channels and infiltration on tilled and untilled loess soil. *Soil Sci.* 119: 242-249.

Fayolle, L. (1979). Consequences of the impact of pollutant on earthworms III. *Laboratory tests Documents Pedozologiques INRA Laboratory Zooecologie et sol*. 1: 34-65.

Fileldson, R.S. (1985). The economic feasibility of earthworms culture on animal waste. *Composting of agril. and other wastes. Ed. Gaser. J.K.R.*: 243-254.

Frapes, G.S. and Sterges, A.J. (1947). Nitrification capacities of Texas soil types and factors which affect nitrification. *Texas Ag. Exp. Sta. Bull.* 693.

Gaddie, R.E. (1980). *News week*. 67-68.

Gaur, A.C. (1983). Enrichment of Urban compost Proc. Int. Symp. Biological Reclaimations and land utilization of Urban Wastes (Eds. Zucconic F. *et al.*) *Institute di Coltivazionic Arboree, Universita degli Studi di Napoli, Naples*: 531-540.

Gaur, A.C. (1992). Bulky organic manures and crop residues. In fertilizers, organic manures, recyclable wastes and biofertilizer. *Ed. H.L.s. Tandon, F.D.C.O., New Delhi.*, pp. 36-53.

Gaur, A.C. and Geeta Singh (1995). Recycling of rural and urban waste through conventional and vermicomposting. In recycling of crop, animal, *Human and Industrial Waste in Agriculture. Ed. H.L.S., Tandon*: 31-49.

Gavrilov, K. (1962). Role of earthworms in the enrichment of soil by biologically active substances. [Voprosy Ekologii Vysshaya Snkila (Moscow)] 7: 34.

Ghabbour, S.I. (1966). *Rev. Eco. Bil. Sci.* 3(2): 259-272.

Ghabbour, S.I. and Imani, M. (1967). The effect of the 5 herbicides on three oligochaete species. *Revised Ecologie and Biologie due sol.* 4: 119-122.

Ghatnekar, S.D. (1994). Industrial waste treatment with worms. *Indian Industrial Sources*. **9**(5): 57-60.

Ghatnekar, S.D.; Kavian, M.F.; Ghatnekar, G.S. and Ghatnekar, M.S. (1998). Management of solid wastes through vermiculture biotechnology. In: Ecotechnology for pollution control and environmental management Ed. R.K. Trivedy and Arvind Kumar. Enviro Media. Karad, pp. 59-67.

Ghilarow, M.S. and Mamajev, B.M. (1966). Uberdie ansteduling von regenwurmern in den artesisch bewasserten oasen der Wurste Kyst-Kum. *Pedologia*. 6: 197-218.

Ghosh, M.; Chattopadhyay, G.N.; Chakravati, P.P. and Bural, K. (1996). A primary study on use of vermicompost in integrated plant nutrient management. A paper presented in National Seminar on developments in Soil Science held at Anand, Gujarat, Oct., 28-Nov. 1, 1996.

Gilman, A.P. and Vardanis, A. (1974). Carboufuran comparastive toxicity and metabolism in the worms. *Lumbricus terrestris* L. and *Ecoenia foetida*. *J. Agril. and Food Chem.* 22: 625-628.

Govindan, V.S. (1998). Vermiculture and vermicomposting. *In: Ecotechnology for Pollution Control and Environment Management* ed. R.K. Trivedy and Arvind Kumar, Enviro Media, Karad, pp. 49-57.

Graff, O. (1971). Stiksoff, Phosphor and Kalium inder wiesenversuch – flache-des solingpojektes. *Ann. Zool. Econ.* 4: 503-512.

Grappelli, K.; Tomati, U. and Galli, E. (1985). Earthworm casting in plant propagation. *Hort. Science*. 20(5): 874-876.

Gunathilagaraj, K. and Sahaya Alfred (1993). Vermiculture, Training division, *Directorate of Extension Education, Tamilnadu Agril. Univ., Coimbatore,* 641 003.

Gupta, M.L.and Ram Sakal (1967). The role of earthworms on the availability of nutrients in grade and cultivated soils. *J. Indian Soc.Soil Sci.* 15(3): 149-151.

Haini, J. and Huhta, V. (1987). Comparision of composts produced from identical wastes by Vermistabilization and convention composting. *Pedobiologia*. 30(2): 137-144.

Hand, P.; Haynes, W.A.; Satcheel, J.E. and Frankland, J.C. (1988). The vermicomposting of cow slurry. *Earthworms in Waste and Environmental Management Ed. C.A. Edwards and E.F. Neuhauser*, pp. 49-63.

Hapse, P.G.; Murkute, S.B. and Zende, N.A. (1993). Effect of vermicompost on sugarcane yield and sugar recovery. *10th annual state level sugarcane development workshop on low cost technology for cane and sugar production organised by V.S.F., Manjari (BK)* on 9-10 July, 1993.

Hatanbe, K.; Ishio Kay and Furuichi, E. (1983). Cultivation of *Elsenia foetida* using dairy waste sludge cake. In *Earthworm ecology from Darwin to Vermiculture*, J.E. Satechell (eds.). Chapman and Hall Ltd., pp. 232-33.

Hopkins, A.R. and Kirk, V.M. (1957). Effect of several insecticides on the English Red Worm. *J. Econ. Ent.* 50: 699-700.

Hopp, H. and Slater, C.S. (1949). The effect of earthworms on the productivity of Agricultural Soil. *J. Agric. Sci.* 78: 325-339.

Hornor, S.G. and Mitchell, M.J. (1981). Effect of the earthworm, *Eisenia foetida* on fluxes of volatile carbon and sulphur compounds from sewage sludge. *Soil Biol. Bochem.* 13: 307-572.

Houser, S. (1993). Distribution and activity of earthworms and contribution to nutrient recycling in alley cropping. *Biol. Fertil. Soils* 15: 16-20.

Inamdar, P.V. (2001). Studies on mineralization of nutrients and humus fractions from vermicompost in lateritic and medium black soils of Konkan. M.Sc. (Agri.) Thesis submitted to Dr. Balasaheb Sawant.Konkan Krishi Vidyapeeth, Dapoli.

INORA (1998). Newsletter of Institute of Natural Organic Agriculture, April, 1998. 2(2): 4.

Ireland, M.P. (1975). The effect of the earthworm *Dendrobaena rubida* on the solubility of lead, zinc and calcium in heavy metal contaminated soils in water. *J. Soil Sci.* 26: 313-318.

Ismail, S.A. (1997). Vermicology. *The biology of earthworms*. Orient Logman, Chennai. pp. 92.

Ismail, S.A. (1998). Vermitech: The Science of earthworm biotechnology Ecotechnology for pollution control and environmental management. Ed. R.K. Trivedy and Arvind Kumar, Enviro media Karad, pp. 109-113.

Ismail, S.A. (1998). Vermitech: The Science of earthworm Biotechnology. *In: Ecotechnology for Pollution Control and Environment Management* ed. R.K. Trivedy and Arvind Kumar, Enviro Media, Karad, pp. 109-113.

Ismail, S.A. and Pramoth, A. (1995). Vermiwash a potent bio-organic liquid "Ferticide" In: Ismail Vermicology. *The biology of earthworms*. Orient Logman, Chennai, p. 92.

Jadhav, A.D. (1996). Effect of FYM and vermicompost on the yield of rice (*Oryza sativa* Linn.) and physico-chemical properties of lateritic soil of Konkan. M.Sc. (Agri.) Thesis, K.K.V., Dapoli.

Jadhav, A.D.; Talashilkar, S.C. and Powar, A.G. (1997). Influence of the conjunctive use of F.Y.M., vermicompost and urea on growth and nutrient uptake in Rice. *J.Maharashtra agric.Univ.* 22(2): 249-250.

Jadhav, S.N.; Patil, G.M.; Kulkarni, B.S. and Patil, R.K. (1998). Growth and yield attributes of mulberry (S-41) as influenced by in situ vermiculture technique. *Advances in Agricultural Research in India.* 9: 119-125.

Jadhav, V.S. (1995). Studies on the physico-chemical changes during humification of organic residues as influenced by earthworms. M.Sc.(Agri.) Thesis submitted to Konkan Krishi Vidyapeeth, Dapoli.

Jadrijevic, D.; Varnero, M.T.; Carrasco, A. and Lopezaliaga, R. (1991). Use of the earthworm *Eisenia foetida* for the decomposition of animal manure II cattle and goat manures. *Advances in Animla Production*. 16(1-2): 189-201.

Jambhekar, H.A. (1990). Effect of vermicompost as a biofertilizer on grape wines. VIIIth Southern Regional Conference on Microbial Inoculent, Pune.

Jambhekar, H.A. (1992). Use of earthworm as a potential source of decompose any organic waste. *National Seminar on Organic Farming. Abstracts Published by M.P.K.V., College of Agriculture, Pune*, pp. 28.

Jansson, S.L.; Hallam, M.J. and Bartholomew, W.V. (1955). Preferential utilization of ammonia over nitrate by micro-organisms during decomposition of oat straw. *Plant and Soil*. 6: 382-390.

Joshi, K.C. (1998). Effect of vermicompost on the growth of forest tree species. Notes of a lecture delivered on Advance training in organic agriculture held at Centre for Advance Studies, Dept. of Soil Sci., JNKVV, Jabalpur 8-22 Sept.

Kachhave, K.G. and Jaishankar, R. (1999). Evaluation of Bio Tech. of vermicomposting as influenced by different methods and organic waste. *64th annual convention of Indian Soc. of Soil Science National Seminar Development in Soil Science*, 1999. Abstract (2). p. 267.

Kadam, R.G. (2000). Effect of vermicompost with and without inorganic fertilizers on yield, quality and mineral nutrition of cowpea-cowpea cropping sequence. M.Sc. (Agri.) Thesis submitted to Dr. Balasaheb Sawant Konkan Krishi Vidyapeeth, Dapoli.

Kale R.D. and Krishnamoorthy, R.V. (1981). Litter preference in the earthworm *Lampito murittii*. *Proc. Indian Acad. Sci. (Anim. Sci.)*. 40(1): 123-128.

Kale, R.D. (1998). *Earthworm ciderella of organic farming*, Prism Books Pvt. Ltd., p. 70.

Kale, R.D. and Bano, K. (1984). Earthworm cultivation and culturing technique. A paper presented to National Seminar on Organic Utilization and Vermicomposting School of Life Science, Sambalpur University, Orissa, 5-8 Dec. p.7.

Kale, R.D. and Bano, K. (1988). Earthworm cultivation and culturing techniques for production of veecomp. 83, E UAS Vee Meal 83 p. UAS, *Mysore J. Agric. Sci*. 22(3): 339-342.

Kale, R.D.; Bano, K.; Secilia, J. and Bagyaraj, D.J. (1989). Do earthworm cause damage to paddy crop. *Mysore J. Agric. Sci*. 23: 270.

Kale, R.D.; Bano, K.I.; Sreenivasa, M.N. and Bagyaraj, D.J. (1987). *South Indian Hort.* 35(5): 433-437.

Kale, R.D.; Mallesh, B.C.; Kuhra Bano and Bagyaraj, D.J. (1992). Influence of vermicompost application on the available macronutrients and selected microbial populations in a paddy field. Soil Biol. *Bio-Chem.* 24(12): 1317-1320.

Kalembasa, S.; Deska, J. and Fiedorow, Z. (1998a). The possibility of utilizing vermicompost in the cultivation of lettuce and tomato. *Ekologiczne aspektly produkcji agrodniczej, Poznan, Poland*, 27: 125-130.

Kalembasa, S.; Deska, J. and Fiedorow, Z. (1998b). The possibility of utilizing vermicompsots in the cultivation of radish and paprika. *Ekologiczene asperktly produkcji ogrodniczej, Poznan, Poland*. **27**: 131-136.

Kang, B.T.; Akinnifesi, F.K. and Pley Sier (1994). effect of agroforestry woody species on earthworm activity and physico-chemical properties of worm casts. *Biol.Fertil. Soils* 18: 193-199.

Kanwar, J.S. (1976). Fertility and properties of submerged rice soils. *Soil Fertility Theory and Practice*. ICAR, New Delhi. pp. 301-317.

Karuna, K.; Patil, C.R.; Narayanswamy and Kale, R.D. (1999). Stimulatory effect of earthworm body fluid (vermiwash) on Crinkle Red Variety of Anthurium andreanum Lind. *Crop Res.* 17(2): 253-257.

Khasnitz, H.G. (1922). *Bot. Arch*. 315-351.

Kiepas, Kokot, A.; Szczech, M. and Fiedorow, Z. (1998). Possibilities of using vermicompost from domestic wastes in ecological plant cultivation. *Ekologiczne aspekty produkcji ogrodniczej, Poznan, Poland.* 27: 137-143.

Kobatke, M. (1954). The antibacterial substances extracted from lower animals. *The earthworms Kekkaby* (Tuberculosis). 29: 60-61.

Krishnamoorthy, R.V. (1984). "A comparative study of worm cast production and nitrogen contribution to soil by earthworm populations from grassland and woodland sites near Bangalore". *A paper presented to National Seminar on organic waste utilization and vermicomposting, School of Life Science, Sambalpur University, Jyoti Vihar, Orissa*, 5-8 Dec.

Krishnamoorthy, R.V. and Vajranabhaiah (1984). An assessment of plant growth promoter level in worm casts. An paper presented to national seminar on organic utilisation and vermicomposting school of life science Sambalpur Univ., Orissa Dec. 5-8, 1984. p. 8.

Kulkarni, B.S.; Nalawadi, U.G. and Giraddi, R.S. (1996). Effect of vermicompost and vermiculture on growth and yield of China aster (*Callistephus chinensis* Nees.) cv. Ostrich Plume Mixed. *South Indian Hort.* 44(1 & 2): 33-35.

Lakhani, K.H. and Satchell, J.E. (1970). Production by *Lumbricus terrestris L.J.Anim.Ecol.* 39: 473-492.

Lavelle, P. (1968). Earthworm activities and the soil system. *Bio.Fertil.Soils* 6: 237-251.

Lebrun, P.; De Medts, A. and Wauthy, G. (1981). Ecotoxicologic comparee of bioactivite detrots insecticides carbamates sultine population experimentate de ver de terre (*Lumbricus hurculeus*). *Pelologia*. 21: 225-235.

Lee, K.E. (1985). Earthworms: Their ecology and relationships with soil and land use. *Academic Press, Austrialia*. p. 411.

Lindwist (1941). Investigation of the importance of some earthworms for demonstration of broad leaf litter and for the structure of mull. *Abs.Biol.* pp. 62-70.

Logsdon, S.D. and Linden, D.R.(1992). Interactions of earthworms with soil physical conditions influencing plant growth. *Soil Sci.* 154(4): 330.

Lozek, O. and Fecenko, J. (1998). Effects of the organomineral fertilizer vermisol special on the quantity and quality of winter wheat yield. Folia – Universitatis. Agriculturae, Stetinensis. *Agricultura*. 1998, No. 72. 185-189.

Lozek, O. and Gracova, A. (1999). The influence of vermisol on the yield and quality of tomatoes. *Acta-Horticulturae-et-Regiotecturae*. 1999. 2(1): 17-19.

Lunt, H.A. and Jacobson, G.M. (1944). The chemical composition of earthworm casts. *Soil Sci*. 58: 367.

Mackay, A.D.; Syres, J.K.; Springett and Gregg, P.E.H. (1962). Plant availability of P in super phosphate and rock phosphate as influenced by earthworm. *In: Ecology of Earth Soil*. **5**: 323-327.

Macrkenzie, B.M. and Dexter, A.R. (1988a).Axial pressures generated by the earthworm. *Aporrectodea rosea*. *Biol. Fertil. Soil* 5: 323-327.

Madhukeshwar *et al.* (1996). Response of tomato seeds and seedlings to different nutrient substrates. In: Earthworm, Cindrella of organic farming (Kale, R.D.). A prism books Pvt.Ltd., 1998, p. 54.

Mahendran, P.P.; Kumar, N. (1997). Effect of organic manure on cabbage cv. Hero (*Brassica oleraceae* Var. Capitata L.). *South Indian Hort.* 45(5-6): 240-243.

Manna, M.C.; Singh, M.; Kundu, S. and Tripathi, A.K. (1994). Decomposition of waste organic materials of farm and city as influenced by vermiculture extended summaries. *National Seminar on Developments in Soil Science on Nov. 28 – Dec. 1, 1994 at Indian Agricultural Research Institute,* New Delhi, 110 012.

Martin, L.W. and Wiggans, C.S. (1959). The tolerance of earthworms to certain insecticides, herbicides and fertilizers. *Oklhoma State University Experiment Station Proceeding Series No. 334.*

Mathukia, R.K.; Ramani, V.B.; Chovatia, P.K. and Joshi, R.L. (1999). effect of integrated nutrient management on yield and quality of sugarcane (*Saccharum officinarum* L.). *Indian Sugar* 48(10): 839-842.

Mba, C.C. (1978). *Z. Pflezenernaehe* Bodenk. 141: 453-478.

Mishra, P.C. (1984). "Soil pollution and earthworm" in Souvenir. A paper presented to *National Seminar on organic waste utilization and vermicomposting. School of Life Science, Sambalpur University, Jyoti Vihar, Orissa, 5-8,* pp. 24-29.

Mitchell, M.J.; Mulligan, R.M.; Hatenstein, R. and Neuhauser, E.F. (1977). Conversion of sludges into 'top soils' by earthworm *Elsenia foetida. J. Environ. Qual.* 9: 373-378.

Mitsui (1955). Inorganic nutrition fertilization and soil amelioration for low land rice. 2nd Ed. Yokendo Press, Tokyo. pp. 107.

Mrinal, Saikia; Rajkhowa, D.J. and Saikia, M. (1998). Effect of planting density and vermicompost on yield of potato raised from seedling tubers. *J. of the Indian Potato Assoc.* 25(3-4): 141-142.

Needha, A.E. (1957). Components of the nitrogenous excreta in the earthworms *Lubricus terrestris* and *Eisenia foetida* (Savigny). *J. of Exp. Bio.* 34: 425-446.

Nethra, N.N.; Jayaprasad, K.V. and Kale, R.D. (1999). China aster (*Callistephus chinensis* (L.) Nees) cultivation using vermicompost as organic ammendment. *Crop Research*, Hissar, 1999 17(2): 209-215.

Neuhauser, E.F.; Kaplan, D.L.; Malecki, M.r. and Hartenstein, R. (1980). Materials supporting weight gain by the earthworm *E. foetida* in waste conversion systems. *Agric. Wastes*. 2: 43-60.

Nielson, R.L. (1951). Earthworm and soil fertility. Proc. N.C. *Grassld.Assoc.* 13: 158-167.

Nielson, R.L. (1965). Presence of plant growth substances in earthworms demonstrated by paper chromatography and the went pea test, *Nature* (London) 208: 1113.

Nijhawan, S.D. and Kanwar, J.S. (1952a). Physico-chemical properties of earthworm casting and their effect on the productivity of soil. *J. Indian Soc. Soil Sci.* 22: 357-372.

Nijhawan, S.D. and Kanwar, J.S. (1952b). Physico-chemical properties of earthworm casting and their effect on the productivity of soil. *J. Indian Soc. Soil Sci.* 22: 575-583.

Palanisamy, S. (1996). Earthworms and plant interactions. In: Notes on National Training on vermiculture prepared by Directorate of Extension Education. *Tamil Nadu Agri.Univ., Coimbatore*, pp. 42-46.

Parle, J.N. (1963a). *J. gen. Microbiol.* 33: 1-11.

Parle, J.N. (1963b). *Ibid.* 31: 13-12.

Patekar, M.T. (1992). Studies of the earthworms in the Konkan region. M.Sc. (Agri.) Thesis submitted to K.K.V., Dapoli.

Patil, M.P.; Hulamani, N.C.; Athani, S.I. and Patil, M.G. (1997). Response of potato (*Solanum tuberosum* L.) cv. Kufri Chandramukhi to integrated nutrient management. *Advances in Agriculture Research in India*. 8: 135-139.

Patil, N.K. (1993). Effect of application of vermicompost and FYM on release of nutrients and their uptake and yield by maize in different textured soils. M.Sc.(Agri.) Thesis submitted to M.P.K.V., Rahuri (Unpublished).

Patil, V.S.; Tambe, R.S. and Pharande, A.L. (2000). Effect of organic matter and potassium on soil properties and yield of wheat. *State level seminar on soil technology and sustainable agriculture souvenir and Abstracts.* January, 18-19, 2000. p. 81.

Piearce, T.G. (1978). *Pedobiologia*. 18: 153-157.

Piearce, T.G.; Oates, K. and Carruthers, W.J. (1990). *J. Zool. Land.* 220: 537-542.

Pincince, A.B.; Donovan, J.F. and Bates, J.E. (1980). Vermicomposting municipal sludge: an economical stabilization alternative. *Sludge*. 3: 26-30.

Ramchandra, Reddy; Reddy, MAN; Reddy, YTN; Reddy, N.S.; Anjanappa, M. and Reddy, R. (1998). Effect of organic sources of NPK on growth and yield of pea (*Pisum sativum*).

Ramesh, P. (1995). Vermiwash promotes crop growth. *Intensive Agriculture*, March, p. 39.

Ranganathan and Christopher (1994). Effect of vermicompost on soil fertility and response of horticultural crops. *Crop Research*, 8(3): 453-456.

Rao, R.K. (1994). Effect of vermicompost on soil properties and biomass production in maize. Thesis submitted to U.A.S., GKVK, Bangalore.

Reddy, B.G. and Reddy, M.S. (1998). Effect of organic manures and nitrogen levels on soil availability nutrients status in maize-soybean cropping system. *J. Indian Soc. Soil Sci*. 46(3): 474-476.

Reddy, N.S.; Latha, K.M.; Jadhav, J. and Nalwade, V.M. (1998). Trace elements and ascorbic acid in coriander of different days of harvest (*Coriandrum sativum*) grown in soil fortified with different fertilizers. *Indian J. Nutrition and Dietetics*. 35(8): 221-226.

Rhee Van, J.A. (1965). *Pl. Soil*. 45-48.

Riffalds, R. and Leviminzi, R. (1983). Preliminary observations on the role of '*Eisenia foetida*' in manure decomposition. *Agrochimica*. 27(2-3): 271-274.

Ruppel, R.F. and Laughlin, C.W. (1977). Toxicity of some soil pesticides to earthworms. *Kanas Ento. Society*. 50: 113-118.

Russel, B.J. (1909). The effect of earthworm on soil productiveness. *J.Agric.Sci.* (England) 3(11): 246-257.

Sarkar, M.C.; Sachder, M.S. and Datta, S.P. (1998). In: Interaction of soil organic matter with nutrients. Soil organic matter and organic residue management. *Bull. of the Indian Soc. Soil Sci*. 19: 90.

Satchell, J.E. (1963). Nitrogen turn over by woodland population of *Lumbricus terrestris*. Soil Organisms. pp. 60-66. Edited by J. Doeksen and J. Vande, Drift. North Holland Publishing Co., Amsterdam.

Satchell, J.E. (1967). Lumbricidae. *Soil Biology*. Academic Press, London.

Satchell, J.E. (1983). Earthworm Ecology – From Darwin to Vermiculture. Chapman and Hall, London.

Senapati, B.K. and Pani, S.C. (1984). Earthworm an underground wonder in Souvenir. A paper presented to *National Seminar on organic waste utilization and vermicomposting. School of Life Science, Sambalpur University, Jyoti Vihar, Orissa, 5-8*, pp. 9-13.

Sharma, N. and Madan, M. (1983). Earthworms for soil health and population control. *J. Sci. Ind. Res*. 42: 575-583.

Sharma, N. and Madan, N. (1988). Effect of various organic wastes alone and earthworms on the total dry matter yield of wheat and maize. *Biological Wastes* 25(1): 33-40.

Shashidhara, G.B.; Basavaraja, P.K.; Basavaraja, R.; Jagadeesha, R.C.; Nadagouda, V.B.; Sadanandan, A.K.; Krishnamurthy, K.S.; Kandiannan, K. and Korikanthimath, V.S. (1998). Effect of organic and inorganic fertilizers on growth and yield of Byadagi chilli. Proceedings of the National Seminar, Madikeri, Karnataka, India, 5-6 October, 1997, 59-61.

Shinde, P.H. and Gawade, D.B. (1992). Effect of application of farm yard manure on the availability of nitrogen, potassium and boron in soil. Proc. of National Seminar on Organic Farming held at College of Agriculture, Pune from April, 18-19, 1992, pp.10-11.

Shinde, P.H.; Naik, R.L.; Nazirkar, R.B.; Kadam, S.C. and Khaire, V.M. (1992). Evaluation of vermicompost. *Proc. of national seminar on organic farming held at College of Agriculture, Pune*, from April, 18-19, 1992. pp. 54-55.

Shrikhande, and Pathak (1951). A comparative study of physico-chemical characteristics of casting of different insects. *Indian J. Agril. Res.* 21: 401-408.

Shrivastava, O.P. (1998). Integrated use of poultry manure and vermicompost with fertilizer. *Indian J.Agric.Chem.* 31(1): 1-12.

Shrivastava, V.M.S. (1984). 'Earthworms as feed of fish, *Mystels vittatus* (Bioch)'. OP. Cit. p. 6.

Sicar, S.S.G.; De, S.S. and Bhowmik, H.D. (1940). In: Interaction of soil organic matter with nutrients. Soil organic matter and organic residue management for sustainable productivity. *Bull. of Indian Soc. Soil Sci.* 19: 90-110.

Snieg, L. Bury, M. (1998). The results of application of bio-preparations in cultivations of potato and winter oil seed rape. Part – I. The influence of selected biopreparations on potato growth and development. Folia – Universitatics. *Agriculturae. Stetinensis, Agricultura.* 1998. 72: 311-317.

Stenersen, J.; Gilman, A. and Varddnis, A. (1973). Carbofuran its toxicity to the metabolism by earthworm (*Lumbricus terrestres). J. of Agril. and Food Chemistry*. 21: 166-171.

Stockdill, S.M.J. (1982). Effects of introduced earthworms on the productivity of NewZealand pastures. *Pedobiologia* 24: 29-35.

Stockdill, S.M.J. (1982). *Pedobiologia*. 24: 29-35.

Stockdill, S.M.J. and Cossens, G.G. (1966). The role of earthworms in pasture production and moisture conservation. *Proceeding of the New Zealand Grassland Association*. 168-183.

Stringer, A. and Wright, M.A. (1973). The effect of benomyl and some related compounds on *Lumbricus terrestris* and other earthworms. *Pesticide Science*. 4: 165-170.

Talashilkar, S.C. (1986). Vermitechnology reprinted from *Khadigramodyog*. pp. 397-403.

Talashilkar, S.C. (1986). Vermitechnology. Reprinted from Kadigramodyog, May, 1986, pp. 397-403.

Talashilkar, S.C. and Powar, A.G. (1998). Vermitechnology for Eco-friendly disposal of waste. *Ecotechnology for pollution control and environment management*. Edited by R.K. Trivedy and Arvind Kumar. Sole Distributors Global Science Publications, Published by Environ Media, 2nd Floor, Rohan Heights, Karad, 415 110. pp. 171-197

Todkari (2001). Effect of vermiwash prepared by two methods on growth characteristics, yield and nutrition of three flowering plants. M.Sc. (Agri.) Thesis submitted to Dr. B.S.K.K.V., Dapoli.

Tomati, O.; Galli, E.; Grappelli, A. and Dilena, G. (1990). Effects of earthworm cast on protein synthesis in radish (*Raphanus sativum*) and lettuce (*Lactuga sativa*) seedlings, *Biol.Fert.Soils*. 9: 1-2.

Tomati, V.; Grapelli, A.; Galli, E. and Rossi, W. (1983). Reftilizer from vermiculture as an option for organic waste recovery. *Agrochimica*. 27: 244-251.

Tomiti, V.; Grapelli, A.; Galli, E. and Rossi, W. (1988). Fertilizer from vermiculture as an option for organic waste recovery. *Agrochimica*. 27: 244-251.

Vadiraj, B.A.; Siddagangaiah; Potty, S.N. (1998). Response of coriander (*Coriandrum sativum* L.) cultivars to graded levels of vermicompost. *J.of Spices and Aromatic Crops*. 7(2): 141-143.

Vadiraj, *et al.* (1996). Comparative account of the establishment of cardamom seedlings in different growth media. In earthworm cindrella of organic farming (Kale, R.D.) - A prism books Pvt.Ltd., 1998, pp. 55.

Van Rhee, J.A. (1969). Inoculation of earthworms in a newly drained polder. *Pedologia*. 9: 128-132.

Van Rhee, J.A. (1971). Some aspects of the productivity of orchards in relation to earthworms activity. *Annal of Zoology and Ecology* (Special publication). 4: 99-108.

Vasanthi, D. and Kumaraswamy, K. (1999). Efficacy of vermicompost to improve soil fertility and rice yield. *J. Indian Soc. Soil Sci.* 47(2): 268-272.

Venkatesh; Patil, P.B.; Patil, C.V.; Giraddi, R.S. (1998). Effect of in situ vermiculture and vermicompost on availability and plant concentration of major nutrients in grape. *Karnataka J. Agricultural Sciences* 11(11): 117-121.

Venkatesh, Patil, P.B., Sudhirkumar, K., Kotikal, Y.K. (1997). Influence of in situ vermiculture and vermicompost on yield and yield attributes of grapes. *Advances in Agricultural Research in India* 8: 53-56.

Waksman, S.A. (1948). Humus (Baltimore: Willams and Wilkins): 95.

Wallwork (1983). Earthworm Biology, Edward Arnold Limited, London, pp. 58.

Waugh, J.M. and Mitchell, M.J. (1981). Effect of the earthworm *Eisenia foetida* on sulphur speciation and decomposition in sewage sludge. *Pedobiologia*. 22: 268-275.

Witcamp, M. (1966). Ecology. 47: 194-201. In Sharma, N. and Madan, M. (1983). Earthworms for soil health and pollution control. *J. Sci. Ind. Res.* 42: 575-583.

Yadav, R.S. (1998). Performance of organic fertilizers in rice based cropping system. M.Sc. (Agri.) Thesis submitted to Konkan Krishi Vidyapeeth (Unpublished).

Zrazhevskii, A.L. (1957). Dozdevye cervi kak faktor plodorodija lesngch pocv. Akad Nauk ukvssr.kiev.

4

Enchytraeid Bioresources

V.B. Lal and G. Tripathi

Introduction

When we talk about annelids, usually we refer earthworms and leeches but there are few other animals which have hardly been notice in Indian soils. Among them enchytraeids is a major group of animals in nature which are very agile and fragile and play an important role in soil formation. With the exception of studies carried out on enchytraeids by Prabhoo (1961, 1964), Thambi and Dash (1973) and Lal *et al.* (1981) in tropical soils of India much of the information on ecological and taxonomic status is yet to be explored.

Enchytraeidae is a family of class Oligochaeta. The enchytraeids occur in terrestrial, littoral and aquatic habitats. They are generally most abundant in moist acid soils with high organic matter content (O'Connor, 1967). They are usually pale-coloured worms. Generally they measure from 10-50 mm in length and anatomically they form a relatively simple and uniform group (Fig. 1). Some 600 species of enchytraeids are now known from different parts of the world, but because of their abundance in the Arctic North temperate zones and sensitiveness to drought, they have been supposed to be of arctic origin. Though phylogenetic analyses of Achaetinae (Enchytraeidae: Oligochaeta) and other bisetate enchytraeids indicate that the achaetines include the earliest species of Enchytraeidae but that Achaetinae is not monophyletic. The earliest species of bisetate enchytraeids now extent are restricted to South America, Africa and India. As this part of

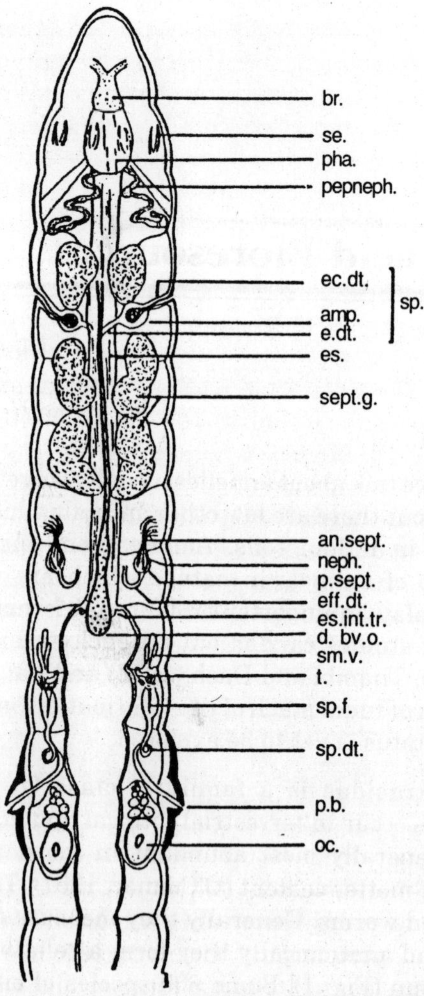

Fig. 1: Tipical morphological characters of an enchytraeids worm amp., ampulla; an. sept., ante-septal; br., brain; d.bv.o. dorsal blood vessel origin; ec.g. actal gland; eff.dt., efferent duct; e.op., ental opening; es., escophagous; es.int.tr., esophageal intestinal transition; m.pha., muscular pharynx; neph., nephridia; oc., oocyte; pha., pharynx; p.b., penial bulb; pepneph., peptonephridia; p.sept., post-setae; sept.g., septal gland; sm.v., seminal vesicla; sp., spermatheca; sp.dt., sperm duct; sp.f., sperm funnel; t., testes.

the enchytraeids lineage is ancestral to other enchytraeids taxa, it is suggested that Enchytraeidae may have arisen in South America or a contiguous southern land mass (Coates, 1989). In Tundra, moorlands, and temperate coniferous forest soils, the enchytraeids from a large percentage of microfaunal biomass. They form some 35.6–92.2% (X = 73.1%) of the total microfaunal biomass in some sites of Barrow, Alaska (Maclean, 1974). Their role in the breakdown of organic matter in sewage beds has been demonstrated by Reynoldson (1939, 1947a, b, 1948). Studies on the role of enchytraeids in litter decomposition in sewage beds, compost pits etc. show that they ingest decomposing plant litter with high microbial content and digest part of it and this partial breakdown of litter makes food material available to other organisms. They also decrease C/N ratio and thus make the litter more acceptable to other organisms and help in releasing nutrients. Their grazing activity by preventing aging of microflora encourage high level of microbial activity. The faecal material excreted by them provide excellent substrate and microhabitat for microbial activity. Thus they are one of the potential decomposer organisms in sewage, beds, compost pits etc. Zachariae (1963, 1964) examined the form and distribution of invertebrate faeces in many soils and concludes that the so called collembolan soils (Kubiena, 1953) have been formed by Enchytraeidae. He claims that enchytraeids take part in soil formation.

The relative importance of an enchytraeids population in a community can be calculated as the percentage of the total annual rate of energy input into the habitat released by the Enchytraeidae population. O'Connor (1967) has calculated that enchytraeids release 11% of the total annual energy input of a Douglas fir plantation in North Wales. The total annual energy input into an aspen forest floor of the Kananaskis area of Alberta, Canada was estimated to be 2360 K Cal/m^2 and the respiratory activity of enchytraeids released about 2.2% of the total annual energy input. The total energy utilized by an enchytraeidae population would amount to about 3-4% of the total energy input if energy expenses for body building and reproduction are included (Dash and Cragg, 1972). Satchell (1971) has calculated the respiration of soil organisms in a broad leaved forest of England and found that the values of fungi, bacteria plus actinomycetes and soil fauna (enchytraeids, nematodes and lumbricids) are 1727, 77 and 361 K cal/m^2 per year respectively. Among the soil fauna, the value for enchytraeids was 167 K cal/m^2 per year. These studies indicate the importance of enchytraeidae in the decomposer system.

Though, the enchytraeids are widely distributed throughout the world, but have not been extensively studied in India possibly due to lack of information on sampling, extraction and taxonomic problems of enchytreaidae in tropical soils.

1. Feeding

A real assessment of the role of enchytraeids in decomposition process could be made only after reviewing their feeding activities. Previous studies on feeding of enchytraeidae (Clark, 1949; Zachariae, 1963; Dash and Cragg, 1972; Dozsa-Farkas, 1973; Standen and Latter, 1977; Dash, 1983; Brockmeyer *et. al.,* 1990; Didden, 1990; Rombke, 1991) showed that they ingested plant litter and grazed over soil fungi. Jegan (1920) states that they ingest plant remains and particles of silica and while burrowing in the soil assist in the subdivision of plant detritus and mix it with the mineral soil. Clark (1949) found that they ingest finely divided plant remains together with considerable quantities of fungal mycelium. Zachariae (1963) has observed that they consume the droppings of litter-feeding collembola, along with all other loose particles of leaf material. He also states that the Enchytraeidae produce crumb-life droppings of finely divided plant remains in which there are no cellulose residues and thus they form a considerable proportion of the moder humus of coniferous forest soils. The Enchytraeidae of sewage beds feed exclusively on algae, fungi and bacteria (Reynoldson, 1939). Enchytraeidae of wrack beds ingest large quantities of decaying seaweed and must play a major part in its decomposition. In laboratory culture they are capable of reducing fresh seaweed to a dark brown amorphous mass of faeces in a short space of time.

Enchytraeus albidus is often aggregated in large numbers in and around the bodies of dead fish and marine birds. Kuhnelt

(1961) states that a liquid given off by the worms liquefies the flesh of the dead animal and that the resulting mess is sucked up by the worms. Enchytraeids found in and around the roots of nematode-infested strawberry plants are able to kill root feeding nematodes. In pot experiments Jegan (1920) found that nematode infestation of strawberry plants could be checked by introducing Enchytraeids to the soil in the early stages of damage.

2. Population Ecology

The soils of the tropical regions of the earth are predominantly oxisols, characteristically low in plant nutrients and organic matter and many with poor physical structure. Their management must be seen against a background of low capital input in economies that lack large resources, coupled with a need to promote sustainable productivity in agriculture and forest ecosystem. Taking it as challenge, the soil biologists are now trying to apply their knowledge of soil fauna to influence soil structure, water and gas exchange, the availability and cycling of plant nutrients, the control of soil losses by erosion. In spite of pioneer studies by Jegan (1920) and Moszynski (1928) who recorded large number of worms from a variety of soil types, this group received little attention. However, the development of reliable quantitative methods for extracting the worms from soil on a large scale (Nielsen, 1952-53; O'Connor, 1995, 1962) opened the way for population studies. Though the seasonal variation does not reveal any systematic variations common to most species or common to all vegetation types in Norwegian forest soil but there is a significant interaction between the vegetation types and sample dates (Abrahamsen, 1972). It means that significant seasonal variation occurs but the peak abundance at different vegetation types does not occur at the same time.

Soil organisms play an important role in releasing the mineral nutrients and thus improve productivity within the ecosystem by decomposition process. So there is a need to asses the relative importance of different groups of pedofauna which help in the breakdown of the organic plant material and thereby enhance the microbial activity of the soil too

(Andren et al. 1988 and Nakamura 1990). The enchytraeids are now well known as secondary decomposers (Dozsa-Frakas, 1978) and they contribute to a great deal to the biomass of the pedoecosystem in a variety of soil types. Climate of the soil to a large extent is dependent on the atmospheric climatic conditions. These not only have direct influence on the nature of the soil but also determines vegetation pattern as well on soil microflora and microfauna. Fluctuations of such abiotic factors may often cause changes in the population density of enchytraeids. The importance of qualitative study of soil fauna have been emphasized by Edwards and Fletcher (1970, 71). Fluctuation in the population density of soil enchytraeids seems to involve several environmental factors. Definitely soil moisture content show a significant positive correlation. In tropical grassland soil of Varanasi in eastern. Uttar Pradesh two peaks in a year have been observed with regard to the population of enchytraeidae (Fig. 2). One of the peak was much pronounced and observed during postmansoon (September-October) and the other, a less distinct one was noted during late winter (February-March).

Seasonal population ecology of enchytraeidae have been worked out in a variety of soil types from different parts of the globe by Nielsen (1955), O'Conner (1957, 1958), Peachey (1962, 1963), Nurimen (1967), Springett *et. al.,* (1970), Abrahamsen (1972), Dash and Cragg (1972), and Didden (1991). With the exception of studies by Thambi and Dash (1973) and Lal (1981), no other ecological information have been reported so far from India.

3. Biodiversity of Enchytraeidae in India

The current and standard monograph of Nielsen and Christensen (1959) with two supplements (1961 and 1963) provide a critical revision of the family Enchytraeidae. They worked mainly on the European species and have revised the Cernosvitov world list. More than six hundred species of Enchytraeidae are known from the different parts of the globe, but only about 5-7% of the total Enchytraeid fauna have been reported from the tropical and sub-tropical regions (Dash, 1983). It is only because of the fact that tropical and sub-tropical

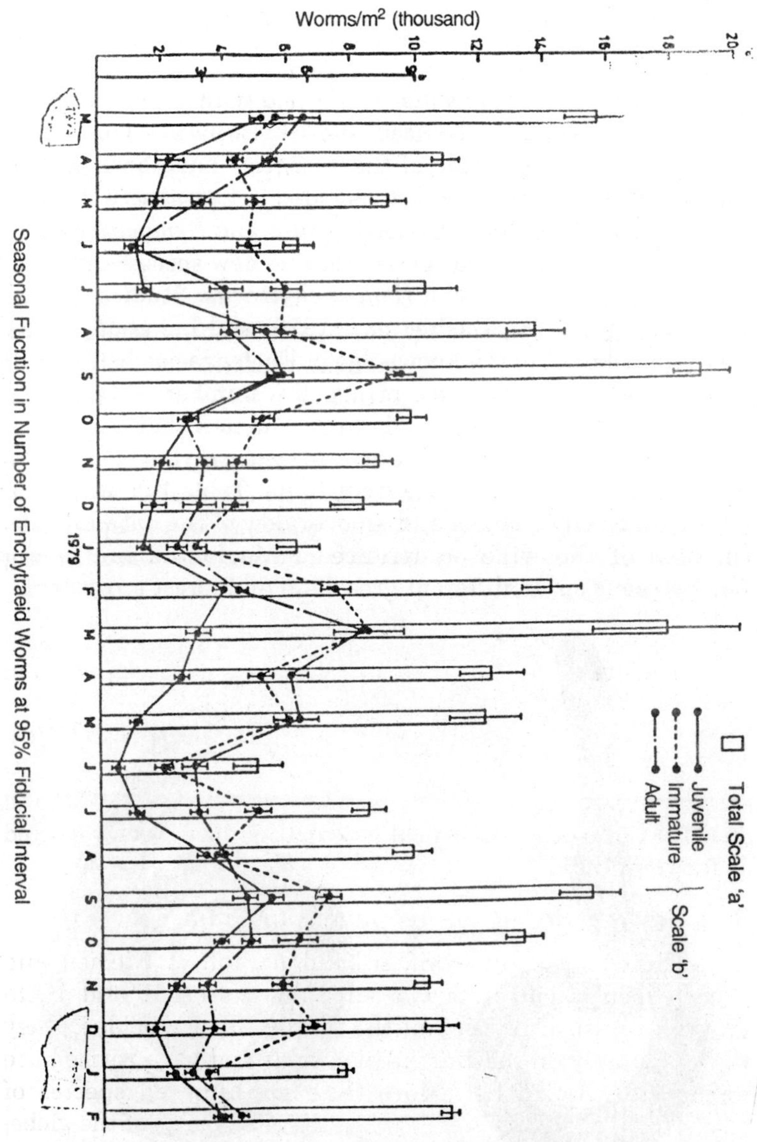

Fig. 2: Seasonal fluctuation in number of enchytraeids worms at 95% fiducial interval.

ecosystems have not been extensively surveyed and the available information shows that these ecosystems harbour many more species.

Genus *Enchytraeus* and *Fridericia* with six species were known from India. In Nielsen and Christensen's monograph (1959) the generic status of these species have been revised and six Indian species are now included in four genera namely– *Enchytraeus, Frideracia, Stephensonilla* and *Hemienchytraeus.* Prabhoo (1961, 1964) has described five new species belonging to the genera *Achaeta* and *Hemienchytraeus.* Studies of Dash and Thambi (1978), Dash *et al.,* (1979) and Lal *et al.,* (1981) have contributed to the knowledge of Enchytraeids biodiversity (Plate 1 and Plate 2) and brings the number of species of tropical Enchytraeidae in India to 31 and now they belong to 9 genera. The genus *Hemifradericia varanensis* (Lal *et al.,* 1981) is reported for the first time from India. Table 1 gives the list of Indian Enchytraeidae and an overview of the data provides an idea of the wide occurrence of Marionina indica and Enchytraeus sp. In different grassland and forest ecosystems.

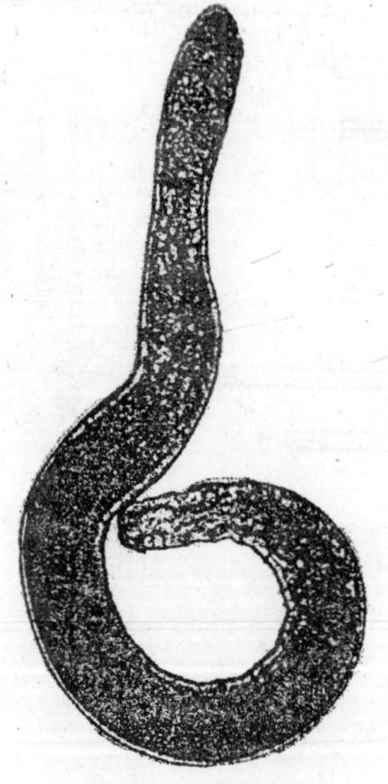

Fridericia indica

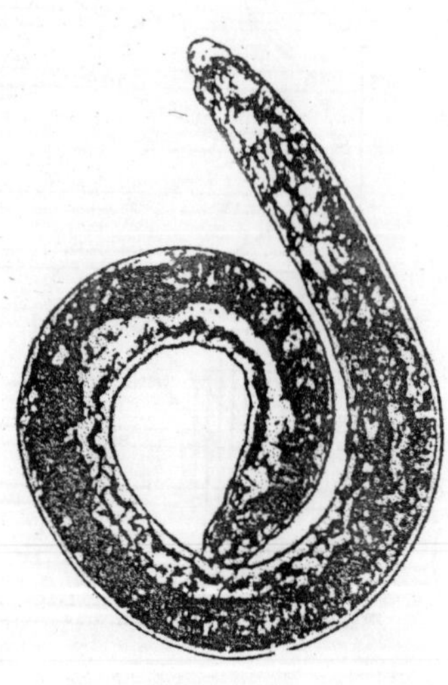

Enchytraeus berhampurosus

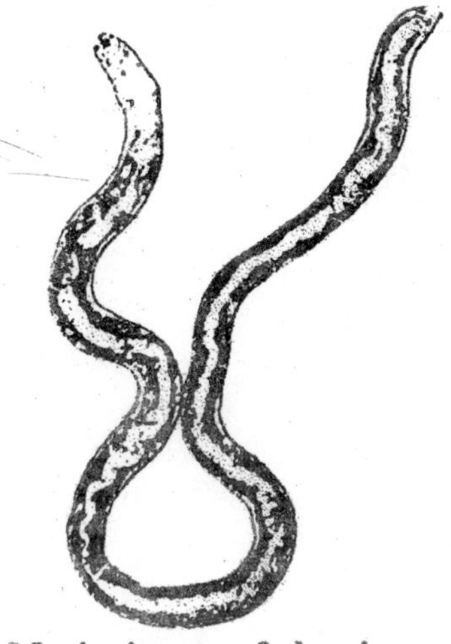

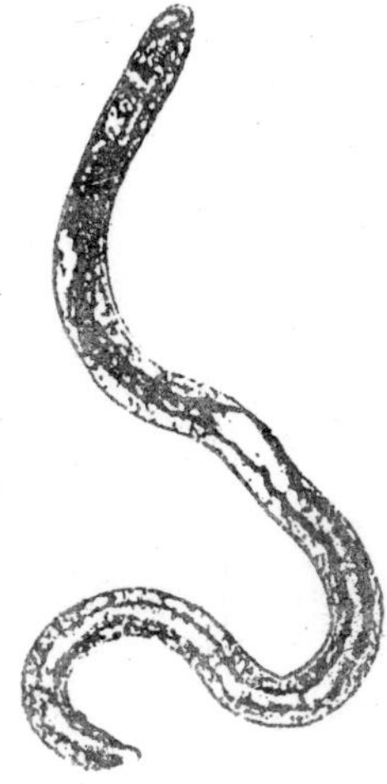

Marionina macfadyeni **Hemienchytraeus khallikotosus**

Plate 1: The enchytraeid species *Fridericia indica, Enchytraeus berhampurosus, Marionina macfadyeni* and *Hemienchytraeus khallikotosus*.

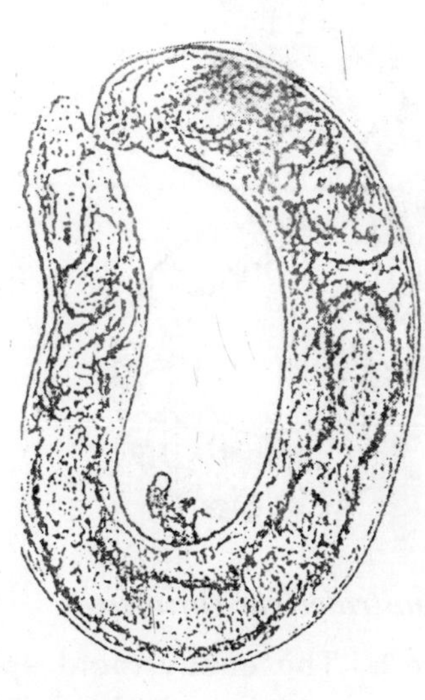

Fridericia dashi **Achaeta nurmineni**

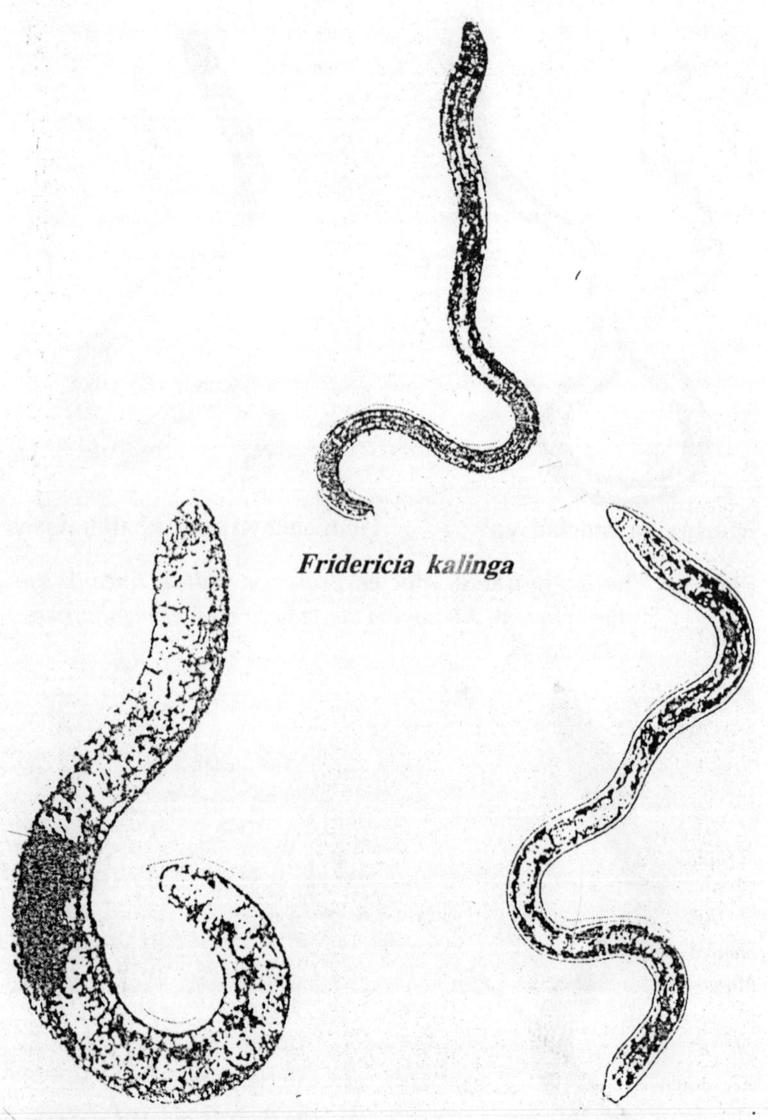

Plate 2: The enchytraeid species *Fridericia dashi, Achaeta nurmineni, Fridericia kalinga, Hemifridericia varanensis* and *Marionina indica.*

Table 1: Taxonomic status of Enchytraeidae in India

Species	*Authority*	*Year*
Achaeta affinia	Nielsen and Christensen	1959
A. christenseni	Prabhoo	1964
A. indica	Prabhoo	1964
A. nurmineni	Dash and Thambi	1978
A. nielseni	Prabhoo	1960
A. segmentata	Prabhoo	1964
Cognettia glandulosa	Michaelsen	1988*
Enchytraeus berhampurosus	Dash and Thambi	1978
E. harurami	Stephenson	1914*
E. indicus	Stephenson	1912*
E. minutus	Nielsen and Christensen	1961
Fridericia bulboides	Nielsen and Christensen	1959
F. bulbosa	Rosa	1887*
F. carmichaeli	Stephenson	1915*
F. dashi	Lal	1981
F. indica	Lal	1981
F. Kalinga	Dash et al.	1978
F. pretoriana	Stephenson	1930*
Hemifridericia varanensis	Lal et al.	1981
Hemienchytraeus bifurcates	Nielsen and Christensen	1959
H. khallikotosus	Dash and Thambi	1975*
H. stephensoni (=*Enchytraeus cavicola* Stephenson, 1924)	Cognettii	1927*
H. theae	Prabhoo	1960
Henlea perpusilla	Friend	1911*
Marionina argentea	Michaelsen	1889*
M. communis	Nielsen and Christensen	1959
M. fulliformis	Nielsen and Christensen	1959
M. indica	Dash and Thambi	1978
M. macfadyeni	Dash and Thambi	1978
M. vesiculata	Nielsen and Christensen	1959
Stephensoniella bercudensis (=*Enchytraeus bercudensis* Stephenson, 1915)	Nielsen and Christensen	1959

* Quoted by Dash and Thambi, 1978.

Acknowledgements

I am very grateful to Prof. M.C. Dash, Chairman, Pollution Control Board, Bhubaneswar, Orissa who provided me the basic training for identification of enchytraeids worms and to Dr. K. Dozsa-Farkas, Department of Systematic Zoology and Ecology, Eotvos Lorand University, Puskin, Budapest, Hungary for providing literatures from time to time.

REFERENCES

Abrahamsen, G. 1972, Ecological study of Enchbytraeidae (Oligochaeta) in Norwegian Coniferous Forest Soils. *Pedobiologia.* 12: 26-82.

Anderen, O., Paustian, K. and Rosswal, T. 1988, Soil biotic interactions in the functioning of agroecosystems. *Agricult., Ecosyst + Envcironment.* 24: 57-67.

Brockmeyer, V., Schmidt, R. and Westheide, W. 1990. Quantitative investigations of the food of two terrestrial enchytraeids species (Oligocheata). *Pedobiologia.* 34: 151-156.

Clark, D.P. 1949. Thesis (unpub.) University of Sydeny. (Quoted by Birch and Clark, 1953).

Coates, K.A. 1989. Phylogeny and origins of Enchytraeidae. *Hydrobiologia.* 180: 17-33.

Dash, M.C. and Cragg, J.B. 1972. Ecology of Enchytreidae (Oligochaeta) in Rocky Mountain Soils. *Pedobiologia.* 12: 323-335.

Dash, M.C. and Thambi, A.V. 1978. A Taxonomis study of Enchytraeidae (Oligochaeta) from grassland soils of southern Orissa, India. *Rev. Ecol. Biol. Sol.* 15: 141-146.

Dash, M.C., Nanda, B.R. and Thambi, A.V. 1979. A new species of Enchytraeidae (Oligochaeta) from forest and grassland soils of Orissa, India. *India Biologist.* 10: 35-37.

Dash, M.C. 1983. Biology of Enchytraeidae (Ologochaeta). Ravandra Pal Singh Gahlot Publ Dehradun, India. 1971 p.

Didden, W.A.M. 1990. Involvement of Enchytraeidse (Oligochaeta) in soil structure evolution in agricultural fields. *Biol. Fert. Soils.* 9: 152-158.

Didden, W.A.H. 1991. Population ecology and functioning of Enchytraeidae in some arable farming systems. *Diss. Univ. Wageningen.* p. 116.

Dozsa-Farkas, K. 1973. Ananeosis, a new phenomenon in the life history of the Enchytraeids (Oligochaeta). *Opusc. Zool. (Budap.)*, 12 (1.2): 43-55.

Dozsa-Farkas, K. 1978. The significance of two Enchytraeidae in the decomposition of hornbean litter in mesophillic deciduous forests of Hungary. *Acta. Zool. Acad. Sci. Hung.* 24: 321-330.

Edwards, C.A. and K.E. Fletcher 1970. Assessment of terrestrial invertebrate populations. In: Methods of Study in Soil Ecology, (ed. J. Phillipson), UNESCO. pp. 57-66.

Edwards, C.A. and Fletcher, K.E. 1971. A comparison of extraction methods for terrestrial Arthropods. Methods of Study in Quantitative Soil Ecology. IBP Handbook No. 18 (Phillipson, J., ed.) Blackwell Sc. Publ. Oxford. pp. 150-185.

Fisher, R.A. and Yates, F. 1963. Statistical Table for Biology, Agricultural and Medical Research, 6th ed., Oliver and Boyd, Edinburgh.

Jegan, G. 1920. Landw. Jb. Schweiz. 34: 55-71.

Kubiena, W.L. 1953. The Soils of Europe. Thomas Murty, London. p. 318.

Kuhnelt, W. 1961. Soil Biology. Faber and faber, London. p. 399.

Lal, V.B. 1981. An ecological study of Enchytraeidae (Oligochaeta) in tropical grassland soil Varanasi, India. Ph.D. thesis. Banaras Hindu University, Varanasi. p. 105.

Lal, V.B.; Singh, J. and Prasad, B. 1981. Hemifridericia varanensis, a new species of Enchytraeidae (Oligochaeta) from tropical soils of eastern Uttar Pradesh, India. *Rev. Ecol. Biol. Sol.* 18 (2): 263-267.

Maclean, S.F., Jr 1974. Primary production, decomposition and the active of soil invertebrates in tundra ecosystems: a hypothesis. In: Soil Organisms and Decompositions in Tundra (ed. A.J. Holding *et al.*) Tundra Biome Steering Committee, Stockholm. pp. 1997-206.

Moszynski, A. 1928. Influence des conditions ecologique sur la distributions des Enchytraeides. *Kosmos J. Soc. Polon. Nat. Kopernik.* 53: 731-766.

Nakamura, Y. 1990. Oribatids and enchytraeids in eco-farmed and conventionally farmed dryland grainfields of central Japan. *Pedobiologia.* 33: 389-398.

Nielson, C.O. 1952-53. Studies on Enchytraeidae I. A. technique for extracting Enchytraeidae from soil samples. *Oikos.* 4: 187-196.

Nielson, C.O. 1955. Studies of Enchytraeidae 5. Factors causing seasonal fluctuations in numbers, *Oikos.* 6: 153-169.

Nielson, C.O. and Christensen, B. 1959. The Enchytraeidae. Critical revision and taxonomy of European species. Natura Jutlandica. 8(9): 1-160.

Nielson, C.O. and Christensen, B. 1961. The Enchytraeidae. Critical revision and taxonomy of European species. *Natural Jutlandica.* Suppl. 1, 10: 1-23.

Nielson. C.O. and Christensen, B. 1963. The Enchytraeidae. Critical revision and taxonomy of European species. *Natura Jutlandica.* Suppl. 2, 10: 1-19.

Nurimen, M. . 1967. Faunistic notes on northern European Enchytraeids (Oligochaeta). *Ibid.* 4: 567-587.

O'Connor, F.B. 1955. Extraction of Enchytraeids worms from a confiferous forest soil. *Nature.* 175: 815-816.

O'Connor, F.B. 1957. An ecological study of the enchytraeids worms from a coniferous forest soil. *Oikos.* 8: 161-169.

O'Connor, F.B. 1958. Age class composition and sexual maturity in the enchytraeids worm population of coniferous forest soil. *Oikos.* 9: 272-281.

O'Connor, F.B. 1962. The extraction of Enchytraeidae from soil. In: Progress in Soil Zoology (ed. P.W. Mirphy), Butterworths, London. pp. 279-285.

O'Connor, F.B. 1967. The Enchytraeids., In: Soil Biology (ed. A. Burges and F. Raw) Academic Press, London, New York. pp. 231-356.

Peachey, J.E. 1962. A comparison of two techniques for extracting Enchytraeidae from moorland soils. In: Progress in Soil Zoology (ed. P.W. Murphy). Butterworths, London. pp. 286-293.

Peachy, J.E. 1963. Studies on Enchytraeidae (Oligochaeta) of moorland soils. *Pedobiologia.* 2: 81-95.

Prabhoo, N.R. 1961. Studies on Indian Enchytraeidae (Oligochaeta: Analida). I. Description of three new species. *J. Zool. Soc. India.* 12: 125-132.

Prabhoo, N.R. 1964. Studies on India Enchytraeidae (Oligochaeta: Analida). II. Description of two new species. *J. Zool. Soc. India.* 16: 82-86.

ˋ lson, T.B. 1939. Enchytraeid worms and the bacteria bed ethod of sewage treatment. *Ann. Appl. Biol.* 26: 138-164.

Reynoldson, T.B. 1947a. An ecological study of enchytraeid worm population of sewage bacteria bed. Field investigation *J. Anim. Ecol.* 16: 26-37.

Beynoldson, T.B. 1947b. An ecological study of enchytraeid worm population of sewage bacteria bed. Laboratory experiments. *Ann. Appl. Biol.* 34: 331-345.

Reynoldson, T.B. 1948. An ecological study of enchytraeids worm population of sewage bacteria beds. Synthesis of field and laboratory data. *J. Anim. Ecol.* 17: 27-38.

Rombke, J. 1991. Estimates of the Enchytraeidae (Oligochaeta, Annalida) contribution to energy flow in the soil system of an beech wood forest. *Biol. Fertil. Soils.* 11: 255-260.

Stachell, J.E. 1971. Feasibility study of an energy budget for Meathop Wood. In: Productivity of Forest Ecosystem (ed. Dyvigneaus). Ecology and conservation 4. Proc. Brussels Symposium, 1969. pp. 619-630.

Springett, J.A., Brittain, J.E. and Springett, B.P. 1970. Vertical movement of Enchytraeidae (Oligochaeta) in moorland soils. *Oikos.* 21: 16-21.

Standen, V. and latter, P.M. 1977. Distribution of a population of Cognettia sphagnetorum (Enchytraeidae) in relation to microhabitats in a blanket bod. *J. Anim. Ecol.* 46: 213-229.

Thambi, A.V. and Dash, M.C. 1973. Seasonal variation in numbers and biomass of Enchytraeidae (Oligochaeta) population in tropical grassland soils from India. *Trop. Ecol.* 14(2): 228-237.

Zachariae, G. 1963. Was Leisten Collembolan for den Waldhumus? In: *Soil Organisms* (ed. J. Doeksen and J. Van der Drift), Amsterdam, pp. 109-123.

Zachariae, G. 1964. Welche Bedentung haben Enchytraeus in waldboden? In: Soil Micromorphology (ed. A. Jongerius). Wlsevier Amsterdam. pp. 57-68.